ORCHIDS OF MALAYA

ORCHIDS OF MALAYA

Description of more than 130 species
with 229 black-and-white
and 26 colour photographs
Including a comprehensive botanic glossary

L. B. SEGERBÄCK

A. A. BALKEMA / ROTTERDAM / BROOKFIELD / 1992

Frontispiece & dust jacket:
Coelogyne kaliana Cribb, a mountain epiphyte (see page 51).

CIP-DATA KONINKLIJKE BIBLIOTHEEK, DEN HAAG

Segerbäck, L. B.

Orchids of Malaya: description of more than 130 species / L. B. Segerbäck. – Rotterdam [etc.]: Balkema. – Ill., photogr.
With botanic glossary, index, ref.
ISBN 90 6191 700 X bound
Subject heading: orchids; Malaya.

Published by
A.A. Balkema, P.O. Box 1675, 3000 BR Rotterdam, Netherlands
A.A. Balkema Publishers, Old Post Road, Brookfield, VT 05036, USA

ISBN 90 6191 700 X

Printed in the Netherlands

Table of contents

Preface

The purpose of this book is to try to let readers share the author's enthusiasm for Malayan wild orchids. This is far from being a complete account of the orchid flora. Some of the species illustrated are quite common, others less so. One must hope that they will all survive the barbaric attacks of the present age, so that they will be there for future generations to enjoy.

Wherever possible, the author has tried to explain the etymology of the generic names. No such attempt has been made in the case of the specific names, however, since these are easily found in a number of reference books.

The author would like to express his sincere thanks to Dr P.J. Cribb and his colleagues at the Royal Botanic Gardens, Kew, England, for their kind help with the identification of many of the species here described, and for providing photographs of a couple of species, and also to his old friends Hugh and Deirdre Barrett for their encouragement and many valuable suggestions on reading the entire manuscript.

Notes on photography

The photographic equipment used for the illustrations in this book included a Rolleiflex SL66 with extension tubes and 80 + 180 mm lenses for studio work, and an Alpa 10d with a 50 mm Macro-Switar lens for field shots. Photographs were taken in the field for the purpose of documenting the habitat and as a safety measure: To get some kind of a picture of a plant, albeit lacking in detail, should the plant have wilted before having been brought back to be photographed under studio-like conditions.

The films used were Agfachrome 50S Professional for colour work, and Agfapan 100 Professional for black-and-white.

In the studio, all colour macro photographs of flowers were made with the use of an electronic flash (it is an advantage to use a flash with a rather low output which allows it to be used closer to the subject, thus lowering the contrast). A reflecting screen was used at all times, as well as various coloured backdrops (most often black velvet). In order to obtain a sufficient depth-of-field, the aperture was set at f = 22 whenever possible.

For the black-and-white studio photographs, the light source was generally halogen lamps. Very often a yellow-green filter proved useful to reduce the contrast between the otherwise too bright flowers and the darker vegetative parts.

Lars Börje Segerbäck

Introduction

The term 'Malaya' as used in this book is not a political entity, but a convenient term for the southern part of the Malay Peninsula, including West (or Peninsular) Malaysia (132 000 km^2) and Singapore (600 km^2). By 'Malaysia' we here mean the country of Malaysia (i.e. peninsular Malaysia and Sabah, and Sarawak, the two Malaysian states on the island of Borneo).

The Malay Peninsula is the long and narrow continuation of a range of mountains extending southwards from Thailand and Burma. Consequently, a considerable part of Malaya is taken up by mountainous and hilly country (dotted in Map 1). The mountains are rather low, the two highest peaks, Gunung Tahan and Gunung Korbu having an altitude of roughly 2180 m.

Along the coast, there are lowlands, the western coastline having a more or less continuous succession of mangrove swamps and mudflats.

The east coast has sandy beaches with extended swampy stretches, but very little mangroves.

Peninsular Malaysia is situated between, roughly, 1.5° and 6.5°N. As a consequence, the climate is tropical with high temperatures and high humidity and very small seasonal variations. The year can be divided into two monsoon periods with abundant rainfall: The northeast monsoon from October through February, and the southwest monsoon from mid-April to mid-October. There is no pronounced dry period, and heavy rains frequently occur outside the monsoon periods. The annual mean rainfall in Malaya is about 2500 mm.

Temperatures are uniformly high throughout the year, varying between 22°C and 32°C on the coast, and between 13°C and 27°C at moderate elevations in the mountains. Humidity remains very high all the time everywhere.

As a result of this hot humid climate, plant life in Malaya is characterized by lush and rich growth – with tropical rain forest as the climax vegetation. The rain forest in Malaya is especially interesting as it is very old – some 130 million years. There are parts of this rain forest that have been growing without interruption since the early Tertiary era, and it is therefore very rich in species. The orchids are, of course, well represented in the rain forest flora. At a rough estimate, the species number about 800, ranging from minute species of *Bulbophyllum* to the gigantic *Grammatophyllum speciosum*, the largest of all orchids. There are few regions in the world with more species than this.

Based on the altitude at which it grows, and on the geographical situation, the rain forest in Malaya can be divided into three main types:

The lowland rain forest

This evergreen forest has a three-layered structure, where the trees in the top story have an average height of about 30 m. A typical genus of trees is *Dipterocarpus*. The ground is covered with a thin layer of decaying leaves, but the plant life is otherwise surprisingly poor on the forest floor. Contrary to common belief, true rain forest is far

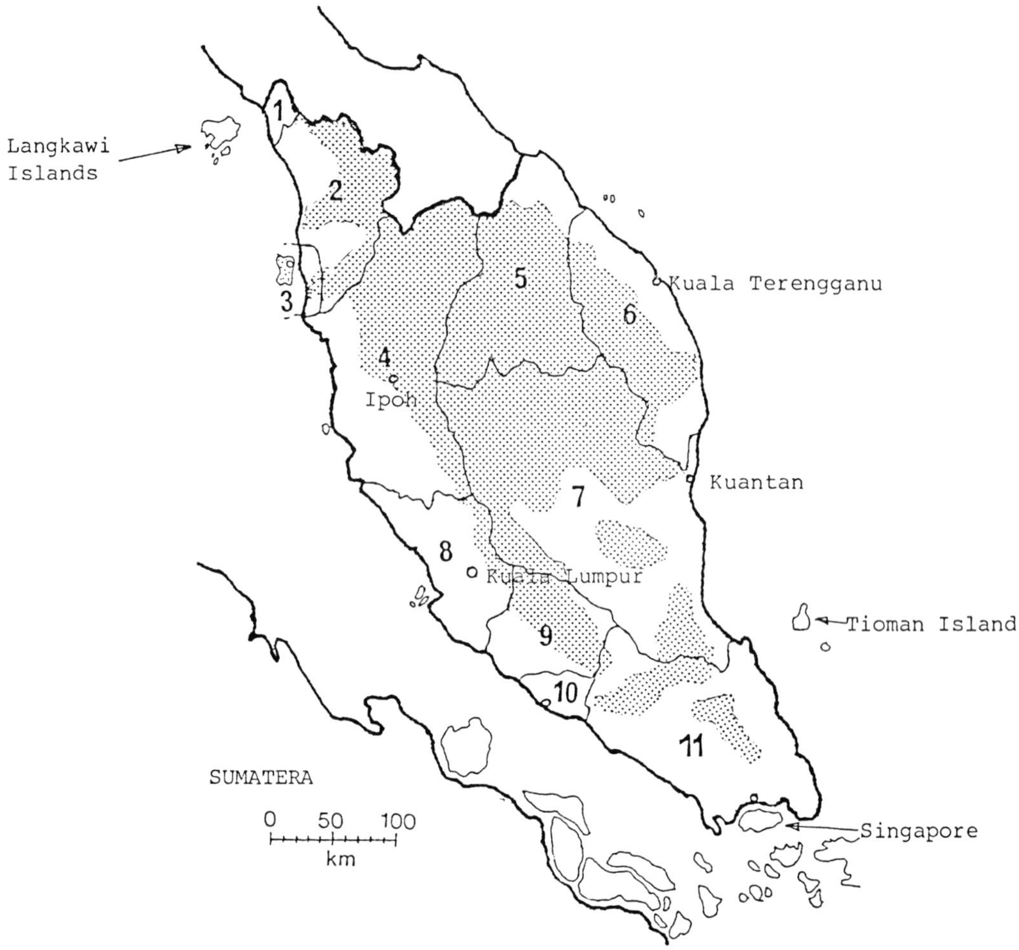

Map 1. Malay Peninsula. The states of Malaya: 1. Perlis, 2. Kedah, 3. Pinang, 4. Perak, 5. Kelantan, 6. Terengganu, 7. Pahang, 8. Selangor, 9. Negri Sembilan, 10. Melaka, 11. Johor.

from being 'impenetrable', it is in fact comparatively easy to walk about in because of the scarcity of vegetation on the ground. Many trees are cauliflorous and/or are buttressed – two typical lowland rain forest features. The number of epiphytes is quite large, including many orchids. As these are growing on the tops of the trees they are often not visible from the ground.

Semi-deciduous monsoon forest

This does not exist in its pure form in Malaya, but in the northern states of Perlis and Kedah the lowland rain forest gradually gets more and more mixed with this type of vegetation. Nowhere in Malaya is there, however, a long dry season of the Indian or African kind. In this forest, there are only two tree stories, and both epiphytes and buttresses are less frequent. There are numerous lianes. It should be noted that although the forest is called deciduous, there are at any time many trees with leaves on them.

Montane rain forest

This is found from an altitude of about 1000 m and upwards. The climate here is much cooler and moister, and the trees much smaller, with practically no buttressing. With increasing altitude, the dipterocarps gradually give way to oaks and, finally, to ericaceous trees. In the most humid type of montane rain forest, the mist or moss forest, the trees have a stunted growth with an average height of 6 m and no buttressing or cauliflory. There is an abundance of epiphytes with many different species, more easily visible and accessible than in other types of forest. The ground flora is richer and more varied than in forests at lower altitudes, including plants that are normally found growing as epiphytes. Often the ground is completely covered with moss.

The Malayan primaeval forest has been much reduced in extent as a result of human interference, and the irreversible destruction of this valuable forest, as of all other tropical rain forests, is getting progressively worse. This is especially true in the case of the lowland forest since it is more easily accessible and lends itself better to modern efficient methods of timber extraction. The tropical rain forest is a precious heritage that belongs to all mankind, and it is indeed tragic when any individual or interested group is considered as owning it and able to dispose of it at will, for short-sighted exploitation purposes. It is our duty, and we also owe it to ourselves, to do what we can to preserve the tropical rain forest so that its animals, plants and people can continue to live in peace and harmony in this terrestrial climax community.

Key to the genera

1. Plants without any green colour; scale-like leaves . . . . *Lecanorchis,* p. 19
1. Plants with green leaves (or, if without leaves, with green stems or green roots)
  2. Two or three fertile stamens, more or less united with column
    3. Lip pouch-shaped; lateral sepals united . . . . *Paphiopedilum,* p. 13
    3. Lip hardly different from petals . . . . *Neuwiedia*, p. 11
  2. Only one fertile stamen, united with column
    4. Very small epiphytic plants without leaves, but with long green roots *Taeniophyllum*, p. 133
    4. Plants with green leaves
      5. Leaves terete
        6. Tall, scrambling plant with large purple flowers *Vanda hookeriana*, p. 155
        6. Smaller, not climbing plants, with small flowers
          7. Plants with rush-like, tufted stems, and small single leaf continuing the line of the stem above the small tufted inflorescence . . . *Ceratostylis,* p. 101
          7. Elongated stems, leafy in lower part, flowers arising from tuft of chaffy bracts in upper part of stem . . . . *Dendrobium* s. Strongyle, p. 69
      5. Leaves flat, or laterally flattened; not terete
        8. Climbing or scrambling plants with elongated stems
          9. Leaves closely spaced, long, but rather narrow; tepals with purple spots; pollinia not stalked . . . . *Dipodium pictum*, p. 113
          9. Leaves well spaced along stem, elliptic to oblong, two stalked pollinia, divided into semi-globular, free halves . . . . *Thrixspermum,* p. 139
        8. Plants without elongated, climbing stems
          10. Terrestrial plants
            11. Base of stem with pseudobulb near ground level
              12. Pseudobulbs 1-leaved
                13. Leaf attached to base of pseudobulb, inflorescence arising from top of pseudobulb . . . . *Liparis*, p. 33
                13. Leaf attached to top of pseudobulb
                  14. Lip joined to whole length of column . . . . *Calanthe,* p. 21
                  14. Lip hinged to base of column . . . . *Plocoglottis,* p. 25
              12. Pseudobulbs with more than 1 leaf
                15. Lip joined to the whole length of column . . . . . . *Calanthe,* p. 21
                15. Lip not joined to column, or joined to its base only
                  16. Lip at top of flower . . . . *Malaxis*, p. 33
                  16. Lip at bottom of flower
                    17. Lip broad, not 3-lobed . . . . *Liparis*, p. 33
                    17. Lip 3-lobed
                      18. Lip with a spur . . . . *Eulophia*, p. 117
                      18. Lip without a spur

19. Midlobe of lip very narrow at base and much broadened at apex *Spathoglottis*, p. 27
19. Midlobe not very narrow at base . . . . . . *Cymbidium,* p. 115

11. Plants without distinct pseudobulbs above ground level
20. Stems creeping on the ground; ends of stems turning up, bearing inflorescence . . . . *Malaxis*, p. 33
20. Stems not creeping on the ground; leafy shoots from underground rootstock, tuber or pseudobulb
21. Stems long and leafy
22. Plants usually with underground tubers; inflorescence terminal; pollinia 2, granular
23. Lobes of stigma not freely extending in front of column
24. Neither stigma-lobes nor anther-cells extending in front; stigma lobes connate with hypochile and staminodes . *Peristylus,* p. 17
24. Base of anther-cells extending from column; rostellum forming broad band above the coherent and not very prominent stigma-lobes . . . . *Pecteilis*, p. 15
23. Stigma-lobes on processes extending from column and not joined to hypochile . . . . *Habenaria*, p. 13
22. Not this combination of characters; no underground tubers
25. Floral bracts close together and overlapping in 2 opposite rows *Bromheadia*, p. 119
25. Bracts otherwise
26. Leaves long and narrow, grasslike; terminal inflorescence with large purple flowers . . . . *Arundina graminifolia*, p. 29
26. Leaves ovate, stiff; large white bracts; white flowers *Dilochia*, p. 31
21. Stems shorter, leaves radical
27. Lip at top of flower, large flowers . . . . *Cryptostylis,* p. 19
27. Lip at bottom of flower
28. Lip without a spur; inflorescence with small, spirally arranged flowers . . . . *Spiranthes*, p. 21
28. Lip with a spur, inflorescence otherwise
29. Lip joined to base of column only; spur short . . *Eulophia,* p. 117
29. Column joined for its whole length to base of lip; spur often long *Calanthe*, p. 21

10. Epiphytic (or sometimes lithophytic) plants
30. Stems with distinct pseudobulbs
31. Pseudobulbs 1-leaved
32. Inflorescence very short, flowers appearing among chaffy bracts at base of leaf . . . . *Dendrobium* s. Desmotrichum, p. 55
32. Inflorescence otherwise
33. Inflorescence from top of pseudobulb
34. One flower open at a time; leaves broad, plicate . . . *Coelogyne,* p. 47
34. Several flowers open at the same time; leaves narrow
35. Sepals hairy on dorsal side . . . . *Eria elata*, p. 79
35. Tepals glabrous

36. Leaves jointed at base; column without lateral projections or wing round the top . . . . *Liparis*, p. 33
36. Leaves not jointed at base; column with lateral projections or wing at the top
37. Narrow arms on sides of column . . . . . . *Dendrochilum*, p. 39
37. No arms on sides of column, but top of column winged
38. Large, conspicuous, persistent, imbricate bracts in 2 rows *Pholidota pallida*, p. 45
38. Large, but well-spaced bracts, falling when flower opens *Chelonistele sulphurea*, p. 47
33. Inflorescence not arising from top of pseudobulb
39. Column with foot, to which the usually tongue-shaped lip is moveably attached . . . . *Bulbophyllum*, p. 83
39. Column without a foot; pseudobulbs narrow, spindle-shaped *Dendrochilum*, p. 39
31. Pseudobulbs with 2 or more leaves
40. Inflorescences from the top or near the top of the pseudobulbs; or at the ends or near the ends of the stems
41. Shortly persistent flowers from groups of chaffy bracts on leafless ends of stems . . . . *Dendrobium crumenatum*, p. 63
41. Inflorescences otherwise
42. Very short inflorescences with a few flowers *Eria leiophylla*, p. 75
42. Long inflorescences with several flowers
43. One flower open at a time
44. Rhachis zigzag, glabrous; flowers large *Coelogyne carnea*, p. 49
44. Rhachis straight, hairy . . . . *Eria pulchella*, p. 75
43. More than 1 flower open at a time
45. Pseudobulbs inconspicuous; small, helmet-shaped flowers *Polystachya concreta*, p. 125
45. Not this combination of characters
46. Pseudobulbs with more than 1 joint . . . . . . *Eria elata*, p. 79
46. Pseudobulbs with 1 joint only
47. Flowers with a spur; pseudobulbs brown, angular *Epigeneium*, p. 51
47. Flowers without a spur; pseudobulbs otherwise
48. Flowers small, sepals <6 mm long
49. Bracts large, imbricate; column winged at apex *Pholidota pallida*, p. 45
49. Bracts small; apex of column without wings *Liparis*, p. 33
48. Flowers large, sepals >12 mm long . . . Coelogyne, p. 47
40. Inflorescences not from top or near top of pseudobulb or stem
50. Small plants with small, densely spaced flowers
51. Pseudobulbs as broad as, or broader than tall; no column-foot *Thelasis*, p. 127
51. Pseudobulbs taller than broad; column with a foot *Phreatia*, p. 129
50. Larger plants; flowers not densely spaced

52. Pseudobulbs 2-leaved; bracts large, falling off as flower opens *Coelogyne*, p. 47
52. Pseudobulbs with more than 2 leaves; bracts inconspicuous *Cymbidium*, p. 115

30. No pseudobulbs, or very inconspicuous pseudobulbs; stems arising directly from rootstock
53. Very large plant with elongated, curved stems, long narrow leaves and 10 cm wide flowers . . . . . . *Grammatophyllum speciosum*, p. 117
53. Much smaller plants without this combination of characters
54. Leaves, stem and inflorescence hairy . . . . *Trichotosia*, p. 71
54. Leaves and stem not hairy
55. Leaves laterally flattened, so that no upper or lower surfaces remain except at the sheathing base
56. Inflorescence a terminal, many-flowered spike . *Oberonia*, p. 37
56. Inflorescence lateral
57. Inflorescence rather dense, with many small flowers open at a time *Octarrhena*, p. 127
57. Inflorescence short, with only 1 or 2 flowers at a time
58. Mentum present *Dendrobium* s. Aporum, Oxystophyllum & Strongyle, p. 63-69
58. No mentum . . . . *Bromheadia*, p. 119
55. Leaves with normal upper and lower surfaces; not laterally flattened
59. Stems usually short, with few leaves; leaves sometimes arising from the top of a short stem, or apparently directly from a creeping rootstock
60. Only 1 leaf, from the top of the stem or from the rootstock
61. Leaf from top of stem, which is often covered by sheaths; flowers arising from a group of chaffy bracts . . . . . . *Ceratostylis*, p. 101
61. Leaf arising from minute pseudobulb (not always apparent) on a creeping rootstock; inflorescence without chaffy bracts *Bulbophyllum*, p. 83
60. Leaves more than 1, rising in a tuft from base of stem, or arranged along a short stem, or from the top of a stem
62. Inflorescence terminal, capitate; stem flattened *Agrostophyllum*, p. 99
62. Inflorescence otherwise
63. Inflorescence a tall spike with many small white flowers arranged in whorls; leaves 2 from the top of a slender stem, covered with sheaths . . . . *Eria longifolia*, p. 77
63. Inflorescence otherwise; only from the end of the stem or very close to it
64. Inflorescences branched; flowers helmet-shaped, not resupinate, crowded at ends of branches *Polystachya concreta*, p. 125
64. Inflorescence not branched; flowers resupinate
65. Bracts rather large, round, closely imbricate, distichous *Bromheadia alticola*, p. 121

65. Bracts otherwise
66. Inflorescences hairy, as long as or longer than leaves; flowers pale yellow with purple anther . *Eria robusta,* p. 77
66. Inflorescences shorter than leaves
67. Flowers white, with purple lip and column . . *Eria,* p. 73
67. Flowers white, with longitudinal purple stripes; lip yellow with purple callus; column white
Eria lamonganensis, p. 81
59. Stems usually elongated, with many leaves, and leafy throughout
68. Inflorescences from the old stems after leaves have fallen
Dendrobium s. Pedilonum, p. 57
68. Inflorescences from leafy stems
69. Inflorescence generally terminal
70. Inflorescence a spike with closely imbricate bracts in 2 rows; flowers opening 1 at a time . . . . *Bromheadia alticola,* p. 121
70. Inflorescence otherwise
71. Leaves small, < 12 mm long
72. Leaves oblong, spreading horizontally in 2 rows; capitate inflorescence among chaffy bracts; pollinia 8
Agrostophyllum bicuspidatum, p. 99
72. Leaves not oblong or spreading horizontally; inflorescence not capitate; pollinia 4 . . . . *Podochilus,* p. 107
71. Leaves larger
73. Inflorescences densely capitate, with short, slender, zigzag branches or many bracts . . . . *Agrostophyllum,* p. 99
73. Inflorescences elongated . . . . *Appendicula,* p. 109
69. Inflorescence generally lateral
74. Flowers 1 or very few, rising directly from the stem
75. Leaves fleshy, flowers without a spur . . . *Adenoncos,* p. 137
75. Leaves thin; flowers with a mentum
Dendrobium s. Distichophyllum, p. 61
74. Inflorescence with a number of ± stalked flowers
76. Leaves very small (< 10 mm long and 5 mm wide); very small white flowers
77. Moss-like plants; flowers with purple markings
Podochilus, p. 107
77. Fleshy leaves; flowers in pairs, no purple markings
Microsaccus, p. 135
76. Leaves larger, generally > 25 mm long
78. Slender inflorescence with a few, shortly persistent flowers open at a time
79. Lip moveably attached to long column-foot; pollinia 2, not completely split . . . . *Pteroceras,* p. 151
79. Lip firmly attached to short column-foot; two pollen-masses, each completely split into ± unequal semi-globular free halves . . . . *Thrixspermum,* p. 139

78. Inflorescence with a swollen rhachis; many small yellow flowers, only a few open at a time
Ascochilopsis myosurus, p. 153
80. Flowers more long-lasting; several open at a time
81. Pollinia 2, each with a hole (porate); middle part of lip forming large sac; apical part of lip fan-shaped, attached to front edge of sac . . . . . *Gastrochilus calceolaris*, p. 147
81. Pollinia 2, completely or incompletely split into halves
82. Pollinia completely divided into ± unequal, semi-globular independent halves
83. No distinct column-foot
84. Spur with central ridge in front of entrance to spur; yellow flowers with brown markings
Acampe rigida, p. 143
84. No ridge in front of entrance to spur; small white flowers . . . . *Schoenorchis*, p. 147
83. Column-foot present; spur with longitudinal septum
Cleisostoma, p. 145
82. Pollinia incompletely cleft or split
85. Column-foot lacking or very indistinct
86. Stipes broad, flowers large . . . . *Vanda*, p. 155
86. Stipes linear, curving forward at apex; flowers small
Uncifera tenuicaulis, p. 149
85. Distinct column-foot
87. Spur insignificant or lacking; lip with bifurcate appendage . . . . *Phalaenopsis*, p. 153
87. Spur from middle part of lip; column-foot longer than column . . . . *Pteroceras*, p. 151

Descriptive part

The huge orchid family, Orchidaceae, can be divided into three sub-families, Apostasioideae, Cypripedioideae and Orchidoideae all of which are represented in Malaya.

The primitive Apostasioideae sub-family is treated as a distinct family, Apostasiaceae, by some authors. Members of the sub-family are in fact different from all other orchids (with the exception of the Cypripedioideae) in having more than one fertile stamen. Therefore, the Cypripedioideae and Apostasioideae sub-families are sometimes treated as a single sub-family, Pleonandrae, as distinct from Monandrae, which includes all other orchids. The Apostasioideae also have, for orchids, unusually regular flowers, whose petals and sepals are more or less identical in size and shape.

SUB-FAMILY APOSTASIOIDEAE

The Apostasioideae are all rather large terrestrials with plicate leaves. The bases of the stamen stalks are joined with the style (see Figs 2 & 3); thus there is not the complete union of style and stamens seen in more typical orchids. The fruit is sometimes a berry. The sub-family includes two genera, both found in Malaya: *Apostasia* and *Neuwiedia*. The genus *Apostasia* (four Malayan species) has a branched inflorescence, and flowers with two fertile stamens; *Neuwiedia* has three fertile stamens, and an unbranched inflorescence. There is, of course, no rostellum.

GENUS NEUWIEDIA Blume 1834

Three species are known from Peninsular Malaysia.

Neuwiedia veratrifolia Bl. [Figs 1, 2, 3 & 4]

Large lowland terrestrial with numerous, stalked leaves. Leaf-blade lanceolate-acute, thin, plicate, to 40 cm long, 6 cm wide. Flowering stalk stiff, to 70 cm long, with leaf-like, long (to 17 cm), ovate-acute stem bracts throughout its length.

Inflorescence: Erect, rather densely many-flowered. Rhachis, bracts and ovary finely hairy. The yellow flowers open very little.

Floral bracts: Conspicuous, narrowly lanceolate-acute, to 5 cm long, 0.6 cm wide.

Dorsal sepals: Lanceolate-acute, 10 mm long, 2.5 mm wide; base and apex green, apex with short point.

Lateral sepals: With longer, nearly 2 mm long points; obliquely obovate-acuminate; yellow, with green base and apex; slightly longer than dorsal sepals. All sepals with fine hairs on dorsal side.

Petals: Obovate-acute, yellow with a green base, 10 mm long, 5 mm wide (flattened); with incurved margins. Midrib prominent, green.

Fig. 1. *Neuwiedia veratrifolia* Bl. August. 1.2×.

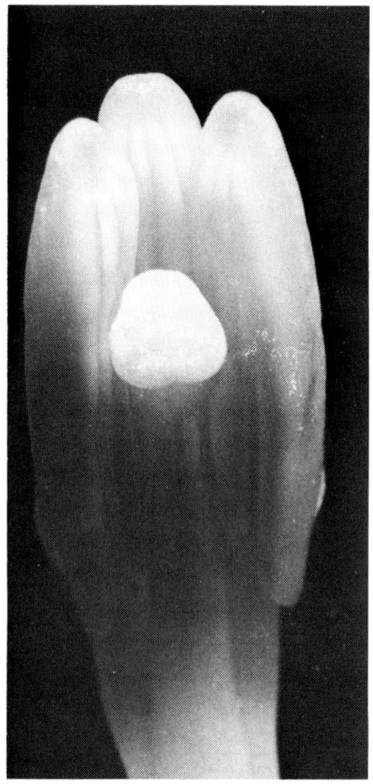

Fig. 2. *Neuwiedia veratrifolia*. Style and three stamina. 4.9×.

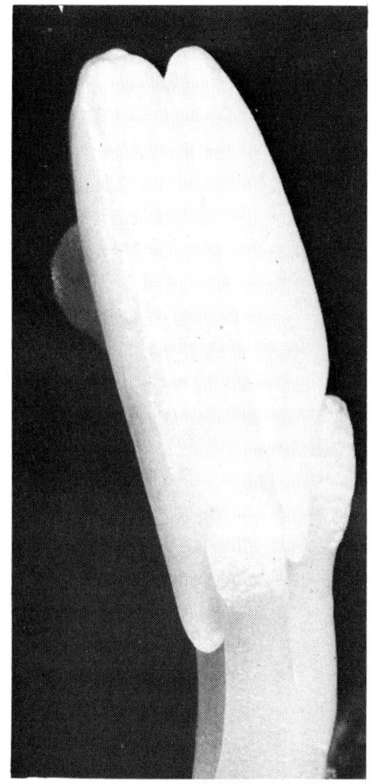

Fig. 3. *Neuwiedia veratrifolia*. Attachment of stamina. 4.9×.

Fig. 4. *Neuwiedia veratrifolia*. 3.3×.

Lip: Oblanceolate-obtuse; similar to petals, but having low median keel, 2 mm wide at base, 0.6 mm wide at the apex.
Ovary: Dark green, 8 mm long, 4 mm wide. The fruit is a capsule.
Distribution: Malaysia, Indonesia, Philippines.

SUB-FAMILY CYPRIPEDIOIDEAE

The Cypripedioideae sub-family has a wider distribution. These orchids are often called 'Slipper Orchids', owing to the form of the lip, and they are represented by the genus *Cypripedium* in the northern parts of Asia, Europe and America. The genera *Selenipedium* and *Phragmipedium* are found in tropical America, and *Paphiopedilum* in tropical Asia. The Asian Slipper Orchids are of great horticultural value, and have been extensively cultivated and hybridized. The name '*Cypripedium*' is still often used for the South Asian Slipper Orchids in horticultural circles.

GENUS PAPHIOPEDILUM Pfitzer, 1886

The peculiar structure of the flower makes the genus very easy to recognize. As mentioned above, there are two fertile stamens, one on each side of the column. A large third stamen, which is sterile, is found at the end of the column. The flowers are generally quite large, and long-lasting. The lateral sepals are joined together, forming a synsepalum behind the slipper-shaped lip. Normally, the dorsal sepal and the petals are widely spreading.

There are five known species of *Paphiopedilum* in Malaya, the best-known species being, perhaps, the very beautiful, white-flowered *Paphiopedilum niveum* from Langkawi Island (Fig. 5). This is now, unfortunately, very rare.

In Greek, 'Paphia' is a name for the goddess Aphrodite, and Paphos is a town in Cyprus celebrated for its temple of Aphrodite. 'Selene' means 'the moon', which played an important role in the cult of the goddess. 'Pedilon' = a shoe, or slipper.

SUB-FAMILY ORCHIDOIDEAE

The orchids with only one stamen, the Monandrae, include among them the great majority of all orchids. They are usually divided into two large groups, Basitonae and Acrotonae. The Basitonae have the base of the anther in contact with the rostellum; this group is not well represented in Malaya and other tropical regions – it includes mainly orchid genera in temperate zones, such as the well-known *Orchis* and *Ophrys*. Instead, most Malayan orchids belong to the Acrotonae-group, where the apex of the anther is in contact with the rostellum.

GENUS HABENARIA Willdenow, 1805

The genus *Habenaria sensu strictu* and some allied genera: *Platanthera*, *Pecteilis*,

Fig. 5. *Paphiopedilum niveum* (Reichenb. f.) Pfitz. Courtesy of Royal Botanic Gardens, Kew.

Fig. 6. *Habenaria angustata* (Bl.) 'O. Kuntze. Alt. 1650 m. September.

Peristylus and a few others, are the only Malayan representatives of the Basitonae-group. These genera are not always well delimited, and some authors prefer to range them all into one large genus, *Habenaria sensu latu*. However even in its narrow sense, the genus *Habenaria* is a very large one (about 800 species), with representatives in all tropical regions of the world. Africa, in particular, has many beautiful *Habenarias*.

Nearly all *Habenarias* are terrestrial. The tuberous root produces an annual shoot, with an unbranched inflorescence. *Habenaria*-flowers are generally small, but quite complicated. The dorsal sepal often forms a hood over the column together with the petals (or with the posterior petal lobes, when the petals are bilobed). The lateral sepals are spreading. The lip is spurred, and its blade often divided into lobes. It is probably the narrow, ribbon-shaped anterior lobes of the petals in many species that are responsible for the name of the genus, from the Latin word 'habena' = a bridle.

The colour of the flowers is generally green or white, but in Malaya there are some species with pink, red or yellow flowers (*H. carnea*, *H. roseata*, *H. rhodocheila*). *Habenaria susannae* (syn. *Pecteilis susannae* (L.) Rafin.) is a beautiful species from northern Malaya, with large, white flowers, spreading petals, and a lip with long, fringed side lobes. The north-western part of Peninsular Malaysia (the States of Perlis and Kedah, including the islands of Langkawi and Pinang), is the best place to look for beautiful species of *Habenaria*. In all, there are some 15 species in Malaya.

Habenaria medioflexa Turrill

One of the beautiful northern species: a highland terrestrial, with a tuber 3 × 1 cm.

Leaves: Numerous, lanceolate-acute, to about 20 cm long and 5 cm wide; base amplexicaul.

Inflorescence: Laxly many-flowered, with several flowers open at the same time.

Bracts: Lanceolate-acute, 20 mm long.

Dorsal sepal: Ovate-acute, hooded, about 6 mm long, green.

Lateral sepals: Ovate-acute, falcate, about 9 mm long, 4 mm wide, green.

Petals: Adnate to dorsal sepal, about 6 mm long, greenish white. Labellum conspicuous, white; mid-lobe linear-acute, 20 mm long, 1 mm wide, deflexed and recurved; side lobes finely divided into feather-like structures, spreading in one plane, more or less semi-circular, 20 × 20 mm. Base of lip with raised edge in front of spur entrance.

Spur: About 3 cm long, more or less flattened; much widened, and with 90° downward bend from the middle; green.

Distribution: Vietnam, Thailand, Malaya.

Habenaria angustata (Bl.) O. Kuntze [Figs 6 & 7]

A highland terrestrial without tubers, but with a thickened rhizome (up to 1 cm wide), and many thick, spreading roots.

Leaves of two kinds: a) large, stalked leaves arising from the rhizome and b) small, sessile leaves on the flowering stalk. The petiolate leaves from the rhizome are elliptic-acute, the blade 7-15 cm long, 3.5-6 cm wide; the petiole is up to 10 cm long. The mid-vein forms a very sharp keel on the lower side of the blade and the petiole. Old leaves are rather stiff, and light green with a dark reticulate pattern. The sessile leaves on the flowering stalk are amplexicaul, often cordate, to 8 cm long, 4 cm wide, and are found in the basal part of the stem. The stem has two basal, scarious sheaths, acute with slightly recurved apex, to 6 cm long. The stem can be

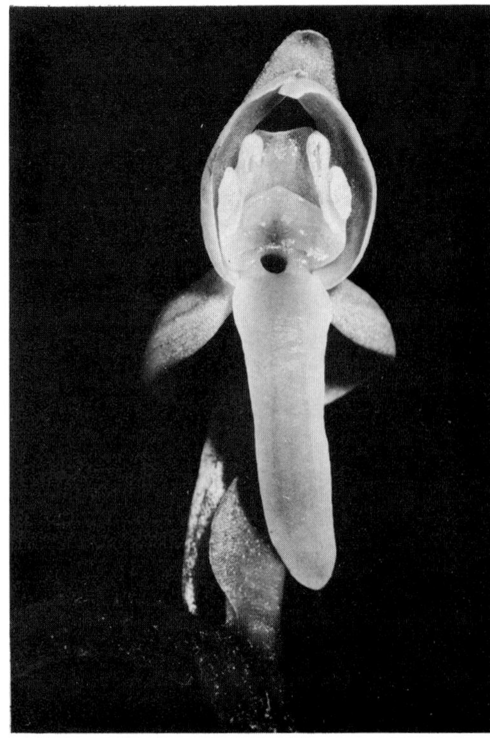

Fig. 7. *Habenaria angustata*. 4.3×.

Fig. 8. *Peristylus lacertiferus* (Lindl.) J.J. Sm. Alt. 600 m. November. 4×.

more than a metre high, and is two-winged, due to the very tightly clasping leaf-sheaths.

Inflorescence: Up to 30 cm long and rather laxly many-flowered, with leaf-like, lanceolate-acute bracts, 13 × 5 mm in the middle of the inflorescence, the size being gradually reduced towards the apex of the inflorescence.

Flowers: Green, and scented.

Pedicel + ovary: 17 mm long.

Dorsal sepal: Ovate-rounded, 6-7 mm long, 5 mm wide, not spreading.

Lateral sepals: Obliquely lanceolate-rounded, 8 mm long, 3 mm wide, spreading or recurved.

Petals: Obliquely lanceolate-subacute, 7 mm long, 3.5 mm wide; not spreading, in contact with dorsal sepal.

Lip: Linear-obtuse, unlobed, 8.5-10 mm long, 2-3 mm wide, with very slightly widened, erect margins at base. The whole lip is sharply deflexed (90°).

Spur: Tubular, curving, and tapering towards the apex, parallel with the ovary, but 1-2 mm longer than the latter. Length of spur 12-15 mm.

Pollinia: Separated by a wide stigma.

Distribution: Malaysia, Indonesia.

GENUS PERISTYLUS Blume, 1825

A genus with many species, mainly found in East Asia and New Guinea. In Malaya, there are some six species. The genus is very closely related to *Habenaria*, and is sometimes included in this genus. In *Peristylus*, the stigmas are sessile, convex, and joined with the lip-base and with the auricles of the column. The anther-processes are either lacking, or very short.

Peristylus lacertiferus (Lindl.) J.J. Sm. [Fig. 8]

A terrestrial, with thickened roots and a tuber. Plant 30 cm high, or more; stem leafy at base only.

Leaves: Broadly lanceolate-acute, amplexicaul, close together, 6.5 cm long, 2.7 cm wide. Stem bracts covering entire length of stem, lanceolate-acute, clasping stem; 17 mm long, 4.5 mm wide.

Inflorescence: Laxly many-flowered (about 25 flowers), rhachis about 10 cm long. Flowers start opening at base of rhachis. Flowers small, white, not much opening.

Floral bracts: Lanceolate-acute, longer than pedicel + ovary; 15 mm long, 4.5 mm wide.

Pedicel + ovary: 10 mm long.

Dorsal sepal: Ovate-subacute, 4 mm long; white with green base.

Lateral sepals: Broadly lanceolate-subacute, 4 mm long; white, with green base and green mid-vein.

Petals: Lanceolate-acute, 3.5 mm long, white. All tepals about 1.5 mm wide.

Lip: Trilobed, 4.5 mm long. Side lobes linear, longer than mid-lobe, deflexed, curving in the same plane as mid-lobe, and ending in an up-turned point. Mid-lobe deflexed, ligulate, 1.5 mm long, 0.6 mm wide. Side lobes pale green, mid-lobe white. Base of lip 1.5 mm wide, with white callus; margins pale green.

Spur: Very short, blunt, sac-like; 0.7 mm long, green.

Column: Short, pale green, with two large, blunt processes in front.

Distribution: India, Burma, Thailand, Sumatra, Java, China, Taiwan.

Fig. 9. *Lecanorchis multiflora* J.J. Sm. September. 1.1×.

Fig. 10. *Cryptostylis arachnites* (Bl.) Hassk. 2.6×.

GROUP ACROTONAE (includes all the following genera)

GENUS LECANORCHIS Blume, 1856

The plants in this small, S.-E. Asian genus (with three Malayan species) are saprophytes: these leafless little orchids with slender stems are found on the floor of lowland rain forests. A distinctive feature in the genus is a small, toothed, cup-like structure below the sepals (the Greek word 'lekanis' means 'a little dish'); it is most easily seen on the seed capsules.

Lecanorchis multiflora J.J. Sm. [Fig. 9]
A saprophyte in lowland rain forest. Scape 35-60 cm long, very slender (1 mm wide), dark purple; with a few acute bracts, 2 mm long.
Inflorescence: At apex of stem, often with flowers and fruits simultaneously.
Floral bracts: Short, decurved, 1 mm long.
Pedicel + ovary: 9 mm long.
Sepals: Lanceolate or oblong, acute, 6 mm long, 1.5 mm wide, pale brownish mauve.
Petals: Oblanceolate-rounded, 7.5 mm long, 2 mm wide, pale brown.
Lip: 9 mm long, trilobed. Mid-lobe spathulate-rounded, 3 mm wide, white, with hairy, pale yellow callus and irregular margins. Base of lip mauve. Side lobes parallel with, and touching, column; pointing forward; the free, distal part 1.5 mm long, white. Base of lip papillose, with low, central keel.
Column: 6 mm long, slender, white; anther white.
Capsules: 15 mm long, 2 mm wide, dark purple; with many ridges, and cup-shaped structure at apex.
Distribution: Malaysia, Indonesia.

GENUS CRYPTOSTYLIS R. Brown, 1810

A small terrestrial genus with some ten species in South Asia, two of which can be found in Peninsular Malaysia. These plants have green leaves: they are not saprophytes. Petals and sepals are very narrow and spreading. The lip is conspicuous, being much wider than the sepals and petals; it narrows gradually to a long point from a wide, concave base, and points upwards, embracing the column (Gk 'kryptos' = hidden; 'stylos' = style).

Cryptostylis arachnites (Bl.) Hassk. [Fig. 10]
This handsome terrestrial is found in lowland rain forests as well as in the cloud forest on exposed mountain peaks. The stalked leaves arise from the rhizome. They are ovate-acute (or acuminate), and light green with a darker reticulate pattern. The glossy leaf-blade is 5.5 to 17 cm long and 3.5-7.5 cm wide; the petiole is 2-15 cm long, and very pale green with purple spots. The height of the plant varies between 10 and 60 cm. The scape is pale green with purple spots, it has many sheaths at the base, and three amplexicaul, lanceolate-acute sheaths higher up.
Inflorescence: Rather laxly many-flowered (about ten flowers). The flowers open singly, or a few at a time, and they have the large lip uppermost.
Bracts: Lanceolate-acute, 1.2-1.5 cm long, green and persistent.

Fig. 11. *Spiranthes sinensis* (Pers.) Ames. Alt. 1600 m. July. 0.9×.

Fig. 12. *Spiranthes sinensis*. 4.6×.

Tepals: All green with incurved margins, which gives them a thread-like appearance. They are also spreading widely. The petals are often wrapped up, and hidden by the dorsal sepal.
Sepals: 17-22 mm long, petals about 10 mm long.
Lip: Lanceolate-acute, and recurved, 15-25 mm long. It makes close contact with the column and folds up lengthwise in mature flowers. The whole lip has fine, velvety hairs. The concave base is pale yellow, changing to brownish purple on the flattened blade. The blade tapers towards the apex which is pale yellow, with dark purple spots.
Distribution: India, Sri Lanka, Cambodia, Vietnam, Thailand, Indonesia, Malaysia, New Caledonia, Solomon Islands, Fiji, Samoa.

GENUS SPIRANTHES L.C. Richard, 1818

A terrestrial genus with a wide distribution, including a large number of species in America, North Africa, Asia, Europe and Australia – New Zealand. The inflorescence with its small, spirally arranged flowers makes the genus easy to identify (Gk 'speira' = screw; 'anthos' = flower). There seems to be one species only in Malaya.

Spiranthes sinensis (Pers.) Ames [Plate 1.1; Figs 11 & 12]
A terrestrial, often seen in open grassland (roadsides) at Fraser's Hill, Cameron Highlands and Genting Highlands. It is further found in Australia and New Zealand, and in many places in Asia, including Japan. The stem is 15-45 cm long and has about three lanceolate-acute leaves, 4-9 cm long, 0.5-1 cm wide, at its base. Higher up, there are three acute bracts, clasping the stem. The rhachis is 5-20 cm long and has many (to about 50), small, spirally arranged flowers with very short pedicels. The floral bracts are lanceolate-longly acute, 5-8 mm long, 1.5 mm wide, and adpressed to the ovary which, together with the rhachis, is finely hairy. Pedicel + ovary 4 mm long. The flowers in Malayan *Spiranthes* are generally white, with a very attractive mauvish-rose tinge.
Sepals: Lanceolate-(sub)acute, 2.5-3 mm long, 1.0-1.5 mm wide, very pale mauve or white; the dorsal sepal darker at the apex.
Petals: Lanceolate to oblong, obtuse to rounded, slightly smaller than sepals; attached to dorsal sepal; white, with mauve apices.
Lip: 3-5 mm long, 2 mm wide (flattened), obovate-retuse, with very irregular margins, apex decurved; blade 2 mm long; white, with pale mauve apex.
Column: Green.
Anther: Brown.
Capsules: Fusiform, slightly falcate, 9 mm long, 4.5 mm wide.

GENUS CALANTHE R. Brown, 1821

Another widely spread terrestrial genus, distributed throughout the Old World tropics, and also in Australia, Japan, China and on Tahiti. Some 20 species are found in Malaya.

These plants are pseudobulbous, although the pseudobulbs are often quite small and easily overlooked when covered by the leaf-bases. Two Malayan species, growing in

Fig. 13. *Calanthe pulchra* (Bl.) Lindl. Alt. 800 m. May. 3.2×.

Fig. 14. *Calanthe angustifolia* (Bl.) Lindl. Alt. 1400 m. July. 1.6×.

the north where there is more seasonal variation in the climate, have large pseudobulbs: *Calanthe vestita* and *Calanthe rubens*. The inflorescence in *Calanthe* is erect and many-flowered. Although not very large, the flowers are often quite attractive, being brightly coloured (Gk 'kalos' = beautiful; 'anthos' = flower). The large, thin leaves are stalked and plicate. A typical feature in *Calanthe* flowers is the tube formed by the union of the lip-base with the sides of the column.

Calanthe pulchra (Bl.) Lindl. [Plate1.2; Fig. 13]

A terrestrial in both lowland and highland forests.

Leaves: Lanceolate, longly acute, plicate, and 30-70 cm long, 4-10 cm wide.

Inflorescence: Erect, 40-60 cm long, pale green, rather densely many-flowered in the apical part, with four large, green stem bracts.

Floral bracts: Fall off before the flower opens; they are lanceolate-acute, pale green, 30-40 mm long and 7 mm wide. The flowers are short-stalked (pedicel up to 11 mm long), they have a waxy appearance and the colour is a beautiful bright golden yellow which, however, rapidly turns into black when the flowers are wilting.

Dorsal sepal: Ovate acute (or apiculate), 16 mm long, 8 mm wide.

Lateral sepals: Narrowly ovate-acute (or apiculate), 16 mm long, 6 mm wide.

Petals: Broadly ovate-apiculate from a narrow base, 15 mm long, 10 mm wide, slightly paler than the sepals.

Lip: Trilobed; side lobes square, spreading, 3 × 3 mm, 7 mm from apex to apex, deep orange. Mid-lobe subspatulate with retuse apex, and apiculus at nodal point; 4 × 4 mm, golden yellow. Two low keels between side lobes. Base of lip projected into spur, 13 mm long (from ovary), apex swollen and bent sharply backwards.

Column: Pale yellow, winged.

Distribution: Malaya, Sumatra, Java.

Calanthe angustifolia (Bl.) Lindl. [Fig. 14]

A mountain species with small pseudobulbs, covered by the leaf-bases.

Leaves: Thin, plicate and very acute, with a stalk 10-20 cm long. The blade is 30-65 cm long, tapering towards apex and base, and 4-7.5 cm wide in the middle. The densely many-flowered scape is 25 cm long, with three sheathing bracts.

Floral bracts: Large, about 4 cm long, fallen when the flowers open. The large, beautiful and fragrant flowers are white with a yellow lip, but turn a deep black very quickly when wilting.

Pedicel + ovary: 20 mm long.

Dorsal sepal: Obovate-acute, 10 mm long, 6 mm wide.

Lateral sepals: Narrowly elliptic-acute, 10 mm long, 5 mm wide.

Petals: Obovate-subacute, 11 mm long, 7 mm wide.

Lip: 14 mm long, trilobed. Mid-lobe obcordate, 6 mm long, 9 mm wide, pale yellow with a darker base, much deflexed. Side lobes ligulate-obtuse, 4 mm long, 2 mm wide, spreading, yellow. Two tooth-like, yellow processes, 1.5 mm long, between side lobes.

Spur: 11 mm long, curving slightly downwards, with widened and slightly flattened apex.

Column and anther: White.

Distribution: Malaya, Sumatra, Java, Philippines.

Fig. 15. *Plocoglottis javanica* Bl. September.

Fig. 16. *Plocoglottis javanica*.

Plate 1

1.1. *Spiranthes sinensis* (Pers.) Ames. 4.4×.
1.2. *Calanthe pulchra* (Bl.) Lindl.
1.3. *Spathoglottis plicata* Bl. 2.5×.
1.4. *Arundina graminifolia* (Don) Hochr., var.α 1.8×.
1.5. *Oberonia costeriana* J.J. Sm. 6.1×.

1.1	1.2	1.3
1.4		1.5

GENUS PLOCOGLOTTIS Blume, 1825

A number of orchids have rather elaborate pollination mechanisms (Many of these have been described by the great Charles Darwin in his *The Various Contrivances...*, which was published in 1877, but is still well worth reading.) In this genus, for instance, most species have a lip with an elastic joint. The lip can be 'primed' by various means, and when it is triggered by the touch of a pollinator, it springs more or less closely into contact with the column. When the pollinator tries to free itself, it is obliged to touch the stigma and the anther. In *Plocoglottis lowii*, the lip is held down by the right-hand lateral sepal which, when touched, releases the lip.

The genus has slender pseudobulbs which are generally one-leaved. The leaf is stalked, rather wide, and plicate. The inflorescence arises from the base of the pseudobulb; it is many-flowered, but only a few flowers are open at a time. The name of the genus seems to be derived from the Greek 'ploke' = plaited, + 'glotta' = a tongue.

Plocoglottis javanica Bl. [Figs 15 & 16]

Lowland terrestrial with thick, numerous, green roots (8 mm wide). Pseudobulbs longly cylindrical, slightly ridged, 6.5-10 cm long, 1.2 cm wide, entirely covered with sheaths, 1-(2-)leaved.

Leaves: Plicate, with five very prominent veins on lower side; elliptical-longly acute, 18-30 cm long, 10 cm wide. Petiole U-shaped in section, with three prominent veins; 4-20 cm long. Scape leafless, with 3-4 acute, scarious stem bracts, to 1.3 cm long. Whole stem dark mauve, puberulent, 40-70 cm long.

Bracts: Triangular-subacuminate, green, 7-8 mm long, puberulent.

Pedicel + ovary: 13 mm long, puberulent.

Tepals: All deep yellow, with irregular red spots, denser toward base.

Dorsal sepal: Lanceolate, tapering to a fine point; 16 mm long, 6 mm wide; dorsal side very sparsely puberulent, ventral side glabrous; base at 90° with ovary, apex decurved.

Lateral sepals: Obliquely lanceolate-acute, 16 mm long, 6 mm wide; dorsal side very sparsely puberulent, ventral side glabrous.

Petals: Basal part ovoid, 4 mm wide, with a few acute teeth on margin of upper part; apical half slightly falcate, almost subulate, yellow, 8 mm long, 2.5 mm wide (at base). Whole petal 15 mm long, glabrous.

Lip: Flexibly attached to short column-foot; blade pale yellow, 7 mm long, 7.5 mm wide; apex wide, with two small, rounded, lateral lobes, and 2.5 mm long, acuminate point in centre. Blade deflexed (at 90° with ovary), when 'primed'. After release nearly parallel with column. Base of lip with small (2.5 mm long), orange sac. Base of blade with two faint purple markings. Side lobes of lip united with column-wings; white, with diagonal purple streak.

Column: Large, 7 mm high, yellow, with two deep purple, rhomboid markings on each side of stigma. Sides of column with large, purple, irregular markings, also covering back of column.

Anther: Yellow.

Distribution: Malaya, Sumatra, Java.

Fig. 17. *Spathoglottis plicata* Bl. Alt. 1000 m. November.

Fig. 18. *Spathoglottis plicata*.

GENUS SPATHOGLOTTIS Blume, 1825

Species of this genus are often cultivated, and can be found growing in gardens in most parts of the tropics. There has been a good deal of hybridization within the genus, which is widely distributed in South Asia, Australia and parts of the Pacific area. In Malaya, there are some six species.

The plants all have rather large, plicate leaves, and tall, slender flowering stalks. The lip is distinctive, being deeply divided into three lobes. The mid-lobe is very narrow at the base, where it is provided with small calli and teeth, and gradually widens toward the apex (Gk 'spathe' = a broad blade; 'glotta' = a tongue). The narrow side lobes are curved upwards.

Spathoglottis plicata Bl. [Plate 1.3; Figs 17 & 18]

This is by far the commonest species, found in the lowlands as well as in the mountains. It is often cultivated, and there are many hybrids and varieties, especially as regards the colour of the flowers. It is a robust species with large, plicate leaves. The inflorescence often looks somewhat untidy, bearing a number of black wilted flowers that remain attached to the capsules. Closer examination of the lip reveals a wealth of beautiful and interesting detail, however.

Leaves: Narrowly elliptical-very acute, tapering and conduplicate towards base; stiff; to 140 cm long, 12 cm wide.

Inflorescence: Tall, many-flowered, with 1-4 flowers open at a time.

Bracts: Broadly ovate-apiculate, much convex, persistent; 2 cm long, 1.2 cm wide; purple, with green apiculus.

Pedicel + ovary: 3.5 cm long.

Dorsal sepal: Elliptic-acute, convex, mid-vein forming low keel on dorsal side; 20 mm long, 10 mm wide.

Lateral sepals: Obliquely elliptic-acute (acuminate), convex, mid-vein forming low keel on dorsal side; 19 mm long, 10 mm wide.

Petals: Elliptic-acute, mid-vein slightly raised on dorsal side; 17 mm long, 12 mm wide. All tepals purple.

Lip: Very conspicuously trilobed, 15 mm long. Front lobe 10 mm long, with a very narrow base (1 mm wide, 5 mm long); widened to blade 7.5 mm wide. Base of front lobe bright yellow, with two longitudinal purple striae; slightly white-hairy; deflexed at point where base of mid-lobe shifts from yellow to purple. Mesochilium yellow, with two suberect, swollen keels, 2.5 mm high, 2 mm wide, bright yellow with purple spots, white-hairy at base and apex. Below keels, the margins of the lip are projected downwards, forming two bluntly triangular, white processes, 1 × 1 mm. Side lobes from base of lip, erect, obliquely spatulate-rounded, 9 mm long, 3.5 mm wide (at apex); base wide, white; 5 mm, narrowing to 2 mm; blade and apex purple. Base of lip white or very pale yellow, with two purple keels.

Column: 14 mm long, slightly curved; pale purple, with darker base and apex. Base of column with two small, white, tooth-like processes in front. Apex of column winged.

Anther: Purple.

Capsules: Cylindric-hexagonal, with three conspicuous ribs (2 mm wide); 35 mm long, 8 mm wide.

Distribution: India, Sri Lanka, Andaman Islands, Burma,

Fig. 19. *Spathoglottis aurea* Lindl. Alt. 1500 m. May. 0.9×.

Fig. 20. *Arundina graminifolia* (Don) Hochr., var. α. Alt. 1200 m. January. 0.6×.

Fig. 21. *Arundina graminifolia*, var. α. 0.9×.

Spathoglottis aurea Lindl. [Fig. 19]

A beautiful mountain species, with large, yellow flowers.

Leaves: Narrowly lanceolate-acute, plicate, up to 150 cm long, 5 cm wide. Petioles up to 25 cm long, purple. Scape up to 130 cm long, purple.

Bracts: Obovate-obtuse, persistent, up to 1.5 cm long, 0.9 cm wide.

Inflorescence: Many-flowered, flowers 7-8 cm across, deep yellow. Tepals widely spreading.

Sepals: Narrowly ovate-subacute, to 40 mm long, 17 mm wide.

Petals: Ovate-obtuse, to 40 mm long, 20 mm wide.

Lip: Trilobed. Side lobes slightly spatulate, 15 mm long, 5 mm wide, base suffused with purple on ventral side. Mid-lobe narrow, slightly widening towards pointed apex; 20 mm long, 4 mm wide (at apex), with one lateral row of purple spots. Two obovoid, thick calli, 5 mm long, erect, at base of mid-lobe, together with two spreading teeth, 4 mm long. Calli yellow, with very small purple dots on apex; base with larger purple markings. Teeth yellow.

Column: Slender and curved, winged, yellow.

Distribution: Malaya, Sumatra, Java.

GENUS ARUNDINA Blume, 1825

Of this terrestrial South Asian genus, only one, very variable species is recognized for Malaya. It is found in all parts of the peninsula, in the lowlands as well as in the mountains.

Arundina graminifolia (Don) Hochr.

This plant is often cultivated, and for good reasons: the flowers are among the largest of all Malayan orchid flowers, and very attractively coloured. Unfortunately, they are not very long-lasting. The plants are tall and stately, and quite easy to grow in the tropics. Consequently, they have been introduced as garden plants in many places, e.g. in Sri Lanka, where they are now well established, and part of the local orchid flora. *Arundina* is sometimes called the Bamboo Orchid, a name that well describes the vegetative habit of the plant, with its elongated stem and long, grass-like leaves. Another local name is 'Tapah Weed'. The generic name comes from the Latin 'harundo' = reed. Two common varieties will be described here:

Var. α (large *Cattleya*-like flowers; frequently cultivated) [Plate 1.4; Figs 20 & 21]

Tall, leafy terrestrial, 90-150 cm high. Stems slender; closely spaced.

Leaves: Numerous, lanceolate-acute, stiff and leathery but not much thickened, 5-25 cm long, 0.6-2 cm wide.

Inflorescence: Many-flowered, often branched. The flowers last for two to three days.

Bracts: Triangular-acute, very stiff, 1-1.5 cm long, 1 cm wide at base. Base of bract clasping stem, apex recurved; pale brown, sticky (from nectaria).

Pedicel + ovary: 2.5 cm long. All tepals spreading.

Sepals: Lanceolate-acute, 4-6 cm long, 1-1.5 cm wide, pale pink.

Petals: Obovate-apiculate, 4-6.5 cm long, 2.5-3.3 cm wide; pink, with paler apices and margins.

Lip: Large, 6-6.5 cm long, trumpet-shaped, base folded round column. Base of lip pale pink, apex and blade purple. Mesochilium with large yellow patch and purple veins.

Fig. 22. *Arundina graminifolia*, var. ß.

Fig. 23. *Arundina graminifolia*, var. ß. Alt. 1100 m. November. 1.1×.

Fig. 24. *Dilochia cantleyi* (Hook. f.) Ridl. Alt. 1550 m. June. 0.2×.

Fig. 25. *Dilochia cantleyi*. 1.5×.

Apex bilobed. Two prominent keels starting 1.5 cm from apex, yellow on blade, changing to purple where yellow patch starts.
Column: Pale pink, with two rounded lateral wings. Front of column purple.
Anther: White.
Capsules: Falcately linear, 5-6 cm long.
Distribution: India, Sri Lanka, China, Burma, Thailand, Malaysia, Indonesia, Indochina, Taiwan, Tahiti.

Var. β (with smaller, less persistent flowers) [Figs 22 & 23]
A terrestrial, 20-100 cm tall, with thin, narrow, grass-like, very acute leaves.
Leaves: To 30 cm long (blade), 1 cm wide (at base); sheath 3.5-4.5 cm long; upper leaves much smaller.
Inflorescence: Terminal, many-flowered, one or two flowers open at a time; flowers shortly persistent. Rhachis to 13 cm long; inflorescence often branching.
Bracts: Spaced about 1 cm on rhachis; triangular-acute, 7 mm long, rather stiff.
Sepals: Lanceolate-acute, 30 mm long, 9 mm wide; white, dorsal side very pale mauve. Laterals not much spreading.
Petals: Obovate-acute, 35 mm long, 16 mm wide; white, dorsal side very pale mauve.
Lip: 35 mm long, 32 mm wide (across side lobes). Side lobes large, but indistinct; overlapping over column; 23 mm long from base of lip; outer side white, with pale mauve margin and darker veins; base of inner side yellow. Base of lip yellow; mesochilium bright yellowish orange. Mid-lobe with two frontal lobes and apiculus in centre; frontal lobes purple. Lip with three distinct keels (yellow), the middle one continuing into apiculus. Basal ends of outer keels with short, narrow processes, pointing toward apex of lip.
Column: White, slightly flushed with pale purple; base of frontal side yellow.
Anther: White.
Capsules: Fusiform, 30 mm long, 7 mm wide, very sharply 6-ribbed; ribs yellowish green, interspaces brownish purple.

GENUS DILOCHIA Lindley, 1830

This genus is limited to South-East Asia, with two species in Malaya. They are terrestrials with long sturdy stems, wide stiff leaves and large, conspicuous bracts.

Dilochia cantleyi (Hook. f.) Ridl. [Figs 24 & 25]
Large terrestrial; stems 1-2 metres long; erect.
Leaves: 5-15 cm long, 2-6 cm wide, ovate-acute, rather stiff; upper side dark green, lower side pale green.
Inflorescence: Terminal, branched. Branches up to 20 cm long, many-flowered. Rhachis and ovary deep purple.
Bracts: Conspicuous, persistent, broadly ovate-acute, much concave, white to cream, up to 20 mm long, 15 mm wide (when flattened).
Dorsal sepal: Lanceolate-acute, 20 mm long, 6 mm wide.
Lateral sepals: Obliquely lanceolate-acute, 20 mm long, 6 mm wide; mid-vein forming distinct keel on dorsal side.
Petals: Lanceolate-acute, 20 mm long, 5 mm wide. All tepals white, flushed with green.

Fig. 26. *Malaxis micrantha* (Hook. f.) O. Kuntze. Alt. 1200 m. March. 3.4×.

Fig. 27. *Liparis caespitosa* (Thouars) Lindl. Alt. 1500 m. July. 0.5×.

Fig. 28. *Liparis caespitosa*. 4.9×.

Lip: Ligulate, trilobed. Side lobes rounded, erect, purple. Mid-lobe rounded, with irregular margin; apical part yellow, with purple markings; basal part greenish white with purple margins; three distinct keels.
Column: Long, slender, arching, with two arms; very pale green, ventral side with purple markings.
Distribution: Malaysia, Indonesia, New Guinea.

GENUS MALAXIS Solander ex Swartz, 1788

Nearly all the plants in this genus are terrestrial, and distributed all over the tropics. There are a few representatives in the northern temperate region as well, e.g. *Malaxis monophylla* (L.) Sw. The genus is large, more than 200 species are known, but of little horticultural interest. The thin, broad leaves (cf. Gk 'malakos' = soft) are generally plicate, and more or less flushed with purple. The inflorescence is terminal, with many, rather small flowers. In these, the lip is always uppermost, it is provided with a toothed apex and has two, comparatively large lobes at the base, near the very short column.

Malaxis micrantha (Hook. f.) O. Kuntze [Fig. 26]
A highland and lowland terrestrial, up to 30 cm high. The lanceolate-acute leaves arise in a tuft from the swollen base of the stem. They are thin, up to 7 cm long and 2.5 cm wide, and dark green, with marked veins forming purple keels on the lower side of the leaf. The petioles and the scape are deep purple, the latter is provided with narrow wings. Floral bracts 4 mm long, lanceolate-acute, sharply deflexed, purple.
Dorsal sepal: Oblong-obtuse, 2.5-3.5 mm long, 1-1.5 mm wide; yellow.
Lateral sepals: Ovate-subacute, 2-3 mm long, 1-2 mm wide; yellow.
Petals: Linear-obtuse, 2.5-3 mm long, 0.3-0.5 mm wide, yellow.
Lip: Sub-orbicular, 2.5-4 mm long and wide, with indistinct side lobes ending in teeth above column; lip yellow, with large purple spot on mesochilium; oblique teeth at apex.
Column: Short, green, with conspicuous stelidia.
Anther: Yellow.

GENUS LIPARIS L.C. Richard, 1818

Another quite large genus with more than 250 known species which are very widely distributed. These plants are terrestrial or epiphytic, with thin, non-jointed, often plicate leaves (section Mollifoliae Rchb. f), or with jointed, coriaceous leaves (section Coriifoliae Rchb. f.). The inflorescence has many small, or rather small flowers, generally with a greenish-yellow colour. The sepals and petals are often narrow and recurved. The lip is much wider, and sharply deflexed near the middle. Together with the long, curved column, these features make the genus easy to recognize. *Liparis* species are very rarely seen in cultivation. Slightly more than 20 Malayan species are known. The name of the genus comes from the Greek 'liparos' = shiny, which probably refers to the appearance of the leaves in the European *Liparis loesellii* (L.) Rich.

Liparis caespitosa (Thouars) Lindl. [Figs 27 & 28]
A highland epiphyte with a very wide distribution (from East Africa to the Pacific).

Fig. 29. *Liparis caespitosa*, var. (Segerbäck 2112). Alt. 1060 m. October. 1.2×.

Fig. 30. *Liparis caespitosa*. 7.5×.

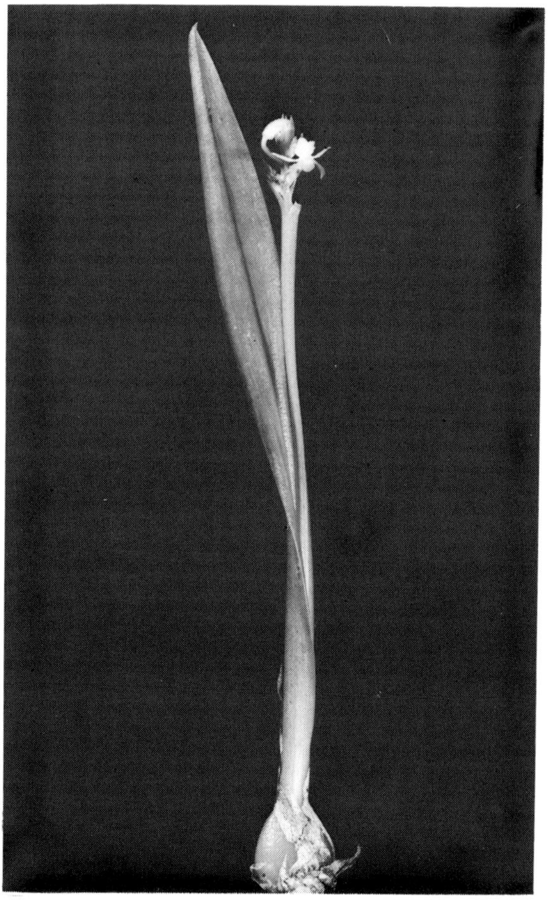

Fig. 31. *Liparis gibbosa* Finet. Alt. 1640 m. November. 0.9×.

Fig. 32. *Liparis gibbosa*. 7.4×.

Pseudobulbs ovoid, closely spaced, 1-leaved, up to 2.5 cm long, 1.3 cm wide.
Leaves: Oblanceolate-acute, thin, not plicate, short-stalked, 10-25 cm long, 1-2 cm wide.
Inflorescence: 7-18 cm long, arising from top of pseudobulb; stalk much flattened and provided with two large, whitish, acute sheaths, to 5 cm long, covering base of inflorescence and leaf. Rhachis to 10 cm long, many-flowered.
Bracts: Narrowly lanceolate-acute, 5-9 mm long.
Flowers: Very pale green, not resupinate. Sepals and paired petals much recurved.
Pedicel + ovary: 4 mm long.
Dorsal sepal: Lanceolate-obtuse, 3 mm long, 0.7-1 mm wide.
Lateral sepals: 3-3.5 mm long, 1-1.2 mm wide, subacute.
Petals: Linear, 3 mm long, 0.2 mm wide.
Lip: 3-4 mm long, 2 mm wide, 'hourglass-shaped': basal part nearly linear, apical part slightly wider, with constriction before apical part. Apex with fine teeth and central apiculus. Lip bent 90° downwards immediately before place of constriction.
Column: With slight S-bend, apex more bent than the rest.
Anther: White.

Liparis caespitosa var. (Segerbäck 2112) [Figs 29 & 30]
A small highland epiphyte. Pseudobulbs ovoid to sub-globular, 10-13 mm high, 8 mm wide, flattened, 1-2-leaved.
Leaves: Oblanceolate-shortly apiculate; thin, not plicate; to 8 cm long (incl. petiole 0.5 cm long), 0.8-1.1 cm wide, with depressed mid-vein.
Inflorescence: Shorter than leaves, much flattened, two-winged; to 6 cm long; rhachis to 3 cm long, many-flowered.
Bracts: Subulate, 5-6 mm long.
Flowers: Very small, pale green, not resupinate.
Pedicel + ovary: 2.5 mm long.
Dorsal sepal: Widening towards slightly pointed apex, 2 mm long, 1.5 mm wide, with depressed mid-vein; apical half sharply deflexed.
Lateral sepals: Elliptic-acute, 2 mm long, 0.8 mm wide, very sharply decurved.
Petals: Linear, 2 mm long, 0.2 mm wide, very sharply decurved.
Lip: Oblong-obtuse, 2 mm long, 0.5 mm wide, very sharply decurved.
Column: Slender, without wings, curving upwards.
Anther: White to pale greenish yellow.
Capsules: Sharply 3-winged, 3 mm long.

Liparis gibbosa Finet [Figs 31 & 32]
A lowland and mountain epiphyte, belonging to a section known as Distichae Ridl. which is easily recognizable from the distichous, imbricate bracts. Pseudobulbs lenticular (ovoid and laterally flattened), more or less erect on sturdy (Ø2-3 mm), branching rhizome; to 1.5 cm high, 1.3 cm wide; single-leaved. Leaf linear-acuminate, to 16 cm long, 1 cm wide.
Inflorescence: Arising from top of pseudobulb; laterally compressed. Stalk very slender, rather stiff, flattened; to 11 cm long, 1.3 mm wide. Rhachis to 4 cm long; many-flowered, but only 1-2 flowers open at a time.
Bracts: Distichous, close together, laterally compressed, lanceolate-acute; to 6 mm long; yellowish-green.
Pedicel + ovary: 8 mm long; orange.

Fig. 33. *Oberonia fungum-olens* Burk. Alt. 1340 m. September. 0.9×.

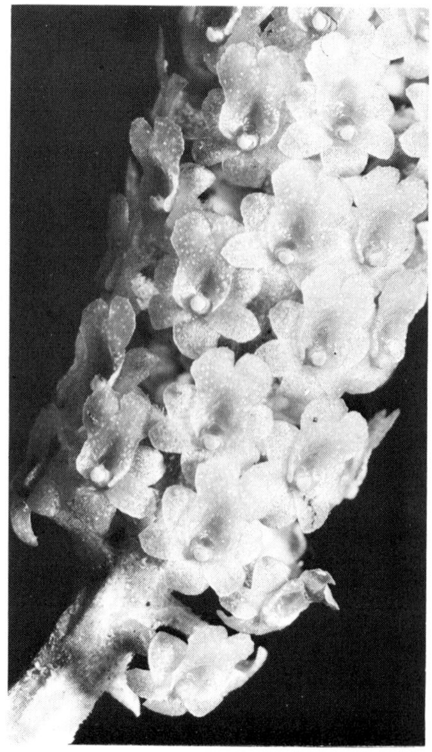

Fig. 34. *Oberonia fungum-olens*. 6.7×.

Fig. 35. *Oberonia costeriana* J.J. Sm. Alt. 1060 m. October. 1.3×.

Fig. 36. *Oberonia costeriana*. 6.7×.

Dorsal sepal: Lanceolate-acute, or acuminate; 5.2 mm long, 2 mm wide; sharply reflexed; orange.
Lateral sepals: Broadly lanceolate-acute (or acuminate); 5.5 mm long, 3 mm wide; slightly oblique; sharply reflexed; orange.
Petals: Very narrowly oblanceolate-obtuse (base linear); 5.5 mm long, 0.8 mm wide; spreading; very pale yellow.
Lip: Triangular-obtuse; orange-coloured; sharply deflexed near base and at apex; 4 mm long (flattened), 3.2 mm wide. Large, white, globular callus, 1.5 mm wide, at base of lip, with 2 small teeth pointing upwards.
Column: White, blunt, with a wide base; 2.5 mm high and wide. Operculum greenish yellow. Column-wings tooth-like, pointing forward.
Distribution: Burma, Laos, Thailand, Malaysia, Indonesia.

GENUS OBERONIA Lindley, 1830

A strictly epiphytic genus with more than 150 species, most of which are found in South-East Asia – with a single species in Africa. The genus is unusually easy to recognize because of the much flattened stems and leaves – all Oberoniae look as if they had just come out of a flower-press. The same feature also occurs in some Dendrobia, Octarrhenae, Bromheadiae and Microsacci, but none of these genera has the typical *Oberonia*-inflorescence, which is terminal, long and spicate, with numerous minute and densely spaced flowers. These are well worth studying with a magnifying glass, as they are often quite attractive. Another unusual feature in the genus is the circumstance that the flowers in the middle of the inflorescence tend to be the first ones to open (this is not true of the African *Oberonia disticha*, though). Some 30 species have been recorded for Malaya. The name of the genus is derived from Oberon, the king of the elves.

Oberonia fungum-olens Burk [Figs 33 & 34]
Highland epiphyte, about 20 cm high. Stem slender, unbranched, flattened, completely covered by imbricate leaf-bases.
Leaves: To 9 cm long, acute, much flattened, to 1 cm wide at base.
Inflorescence: Terminal, spicate, 9-12 cm long, very densely many-flowered, but flowers more sparsely spaced towards apex. Rhachis to 8 cm long, densely hairy and with many narrow, longitudinal ridges; 6 mm wide across flowers.
Bracts: 1.5 mm long, acute and with hairy base.
Flowers: In rather regular whorls, not resupinate.
Pedicel + ovary: Hairy, 1.5 mm long.
Sepals: Ovate-rounded, recurved, slightly less than 1 mm long and wide, pale mauve.
Petals: Slightly narrower than sepals, recurved, very pale yellow.
Lip: Straight, at 90° angle with ovary; 2 mm long, 1.3 mm wide, very indistinctly trilobed. Mid-lobe obcordate; side lobes very small, rounded, pointing at, and nearly in contact with column. Base of lip concave. Whole lip pale yellow.
Column: Short, yellowish mauve.
Anther: Pale yellow. With faint, fungus-like smell.

Oberonia costeriana J.J. Sm. [Plate 1.5; Figs 35 & 36]
Highland epiphyte with very long stem, to 1 cm long.

Fig. 38. *Dendrochilum odoratum*. Alt. 1400 m. September. 3.5×.

Fig. 37. *Dendrochilum odoratum* (Ridl.) J.J. Sm. Pseudobulbs.

Fig. 39. *Dendrochilum linearifolium* Hook. f. Alt. 1650 m. January. 1.1×.

Fig. 40. *Dendrochilum linearifolium*. 3.7×.

Leaves: Much flattened, in one plane, lanceolate-acute (or linear – very acute), falcate; 9.5 cm long. 1.1 cm wide (lanceolate leaves), rather pale green. Linear leaves on some specimens can be up to 12 cm long, 0.6 cm wide.
Inflorescence: Slightly arching, to 13 cm long; rhachis to 10 cm long, many-flowered, flowers start opening 2 cm below apex. Rhachis and stalk pale yellowish brown.
Bracts: Subulate, pale yellow, 4 mm long, conspicuous when flowers in bud.
Flowers: In regular whorls, not resupinate, very small – less than 3 mm between apices of lip and dorsal sepal.
Pedicel + ovary: 1.5 mm long.
Sepals: Ovate-acute, about 1 mm long, pale brick-coloured; laterals often slightly wider than dorsal sepal, and reflexed.
Petals: Oblong to obovate, rounded, 1-1.5 mm long; pale brick-coloured, spreading.
Lip: Triangular, 2 mm long, 1.5 mm wide; indistinctly trilobed. Blade thickened, papillose, brick-red. Side lobes flattened; in line with, and close to blade; pointed, about 1 mm long from base of lip to end of point; brick-red, with yellow, wavy margin. Mid-lobe bifurcate into two oblong or ovate, rounded, more or less up-curved, parallel (sometimes convergent and indistinct), yellow teeth, 0.5 mm long.
Column: Short, brick-red.
Anther: Pale yellow.
Distribution: The species is previously known from Java and Sumatra; the present collection is a new record for Malaya.

GENUS DENDROCHILUM Blume, 1825

This large, but not very well-known genus is exclusively South-East Asian. Malaya has at least 12 out of a total of more than 100 species. The generic name is apparently derived from Gk 'dendron' = a tree and 'kheilos' = a lip. The plants are pseudobulbous, each pseudobulb carrying one leaf. The inflorescence is very slender and often quite attractive in spite of the small size of the flowers, as it is many-flowered, long and gracefully arching, and often scented. At the apex of the column there is a wing round the anther. Further down, there is a separate pair of wings, or arms, on the sides of the column.

Dendrochilum odoratum (Ridl.) J.J. Sm. [Figs 37 & 38]
A highland epiphyte with long, smooth and slender pseudobulbs, which are conical and up to 9.5 cm long, 1 cm wide. The leaf is stalked and rather stiff. The leaf-blade is oblanceolate-acute, to 25 cm long and 3.5 cm wide; the petiole is 1-3 cm long. The inflorescence is synanthous, about as long as the leaf, and densely many-flowered. Scape very slender, 15 cm long. Rhachis 10 cm long.
Bracts: Lanceolate-subapiculate, pale green, 3 mm long.
Flowers: Faintly scented.
Pedicel + ovary: 2 mm long.
Tepals: All green.
Sepals: Lanceolate-acute, 7 mm long, 2 mm wide, not much spreading.
Petals: Lanceolate-acute, 6 mm long, 1 mm wide.
Lip: Unlobed, but margins in basal part slightly upturned. Blade triangular-subacute, curving downwards. Two low keels from apex to base, darker in colour than the rest

Fig. 41. *Dendrochilum gracile* (Hook. f.) J.J. Sm. Alt. 1600 m. March. 0.7×.

Fig. 42. *Dendrochilum gracile*. 6.7×.

Fig. 43. *Dendrochilum gramineum* (Ridl.) Holttum. Alt. 1650 m. February. 0.7×.

Fig. 44. *Dendrochilum gramineum*. 4.7×.

of the lip, which is very pale green. Length of lip 6 mm; width in apical part 1.5 mm.

Column: Short, pale green, with falcate, very acute appendages as high as the anther. Column-wings with irregular teeth. Rostellum pale orange.

Dendrochilum carnosum (Ridl.) Holttum [Plate 2.1]

A highland epiphyte with a preference for rather exposed situations. The long, trailing raceme is often free from the substratum. Pseudobulbs ovoid, 2-2.5 cm long, 1 cm wide. Leaf 6-15 cm long, 1.5-3 cm wide, lanceolate-obtuse, with petiole 2.5-3.5 cm long. Inflorescence from top of pseudobulb, many-flowered. Flowers faintly scented. Scape very slender. Bracts obovate-obtuse, 5 mm long.

Sepals: Lanceolate-acute, 8 mm long, 3 mm wide, green.

Petals: Lanceolate-acute, very slightly shorter than sepals, green.

Lip: Trilobed, 7 mm long. Mid-lobe broadly ovate-acute brown, with upturned, green margins; apex decurved, green. Mesochilium with two parallel brown keels, centre green. Side lobes rhomboid, toothed, brown. Base of lip green.

Column: With curved, pointed white arms below winged, green apex. Anther yellow.

Dendrochilum linearifolium Hook. f. [Figs 39 & 40]

A small attractive mountain epiphyte found on exposed ridges. It has ovoid to sub-globular, closely spaced pseudobulbs, 10 mm high, 5 to 10 mm wide. The leaf is narrowly oblong to linear, obtuse, thin but rather stiff, 30-65 mm long, 4-8 mm wide. The many-flowered inflorescence appears at the top of the pseudobulb. It has a very slender stalk and an arching rhachis. The small bracts are lanceolate-acute. A variety with smaller, pale green flowers and a lip with brown keels flowers in April.

Sepals: Lanceolate-acute, 7 mm long, 4 mm wide, white.

Petals: Ovate-acute, 7 mm long, 5 mm wide, white.

Lip: Trilobed. Side lobes small, erect; mid-lobe ligulate, with two well-defined brown or mauve keels. Rest of lip white.

Column: Very long and slender, with two incurved processes; white.

Dendrochilum gracile (Hook. f.) J.J. Sm. [Figs 41 & 42]

Highland epiphyte with longly ovate pseudobulbs, 10-50 mm long, 5-9 mm wide. Leaf-blade lanceolate-acute, 4-15 cm long, 1.5-3 cm wide. Petiole 5-15 mm long. Inflorescence from top of pseudobulb, 15-20 cm long, many-flowered; stalk very slender. Flowers fragrant.

Bracts: Ovate-acute, reddish, 2.5 mm long.

Tepals: Lanceolate-acute, 5-6 mm long (petals slightly shorter than sepals), 1.5 mm wide, pale green.

Lip: Distinctly trilobed, 3-5 mm long, very pale green. Middle lobe sub-orbicular, acute; apex decurved. Mesochilium with two dark brown keels, changing to green on hypochilium. Side lobes subulate-falcate, short.

Column: Winged at apex, and with two long, lanceolate stelidia.

Distribution: Malaya, Java.

Dendrochilum gramineum (Ridl.) Holttum [Figs 43 & 44]

A highland epiphyte. Pseudobulbs conical, 1-2.5 cm high, up to 1 cm wide.

Leaf: Lanceolate-acute from narrow base, with petiole 5-6 cm long. Blade 12-21 cm long, 1-1.6 cm wide.

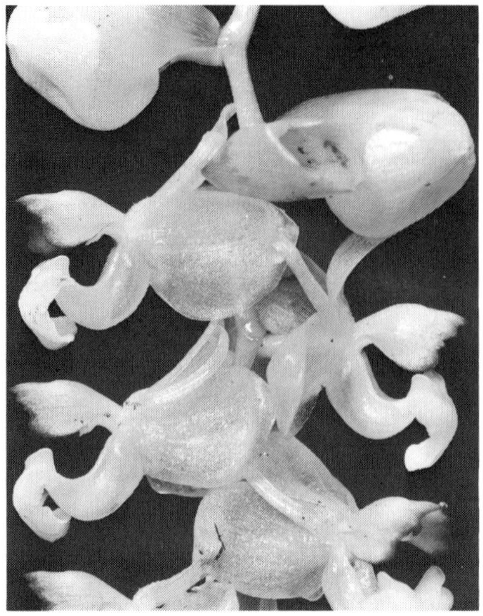

Fig. 46. *Pholidota gibbosa*. 2.8×.

Fig. 47. *Pholidota gibbosa*. 2.6×.

Fig. 45. *Pholidota gibbosa* (Bl.) de Vr. Alt. 1340 m. September. 0.5×.

Fig. 48. *Pholidota parviflora* Hook. f. Alt. 1650 m. June. 1.1×.

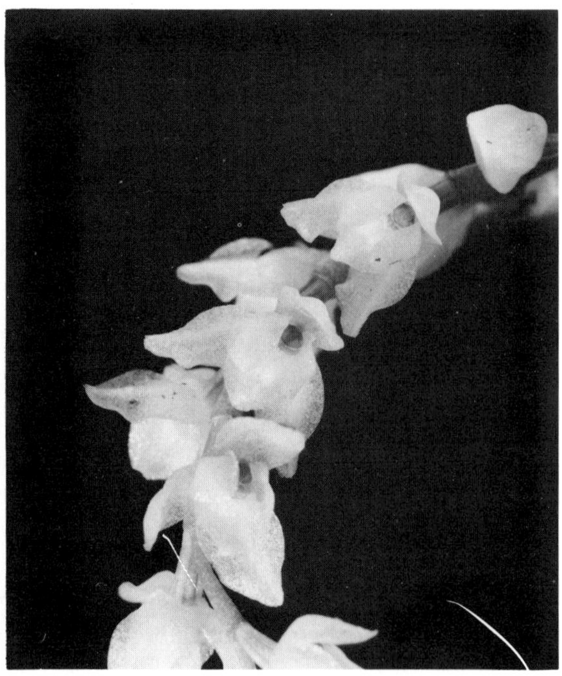

Fig. 49. *Pholidota parviflora*. 4.1×.

Inflorescence: From top of pseudobulb, with very slender stalk; 20-35 cm long, rhachis 5-15 cm long, arching, many-flowered; all flowers open at the same time, and fragrant.
Bracts: Small, scarious, lanceolate-acute, 6 mm long, 1 mm wide.
Tepals: Lanceolate-acute, pale yellowish green, spreading.
Sepals: 6 mm long, 1.5 mm wide.
Petals: 5 mm long, 1 mm wide.
Lip: Ligulate, attached to base of column, with very small side lobes and two narrow keels diverging slightly towards column; whole lip 3-4 mm long, 1 mm wide, very pale green, with two falcate stelidia.

GENUS PHOLIDOTA Lindley ex Hooker f., 1825

A small South Asian-Australian epiphytic genus with one- or two-leaved pseudobulbs and a slender inflorescence. The genus is related to *Dendrochilum*, but lacks the lateral column-arms of this genus. Instead, the apical column-wing is often quite large. The lip has a sac at its base, which is quite distinct from the blade. The bracts are large, concave, and often conspicuous (hence probably the generic name: from Gk 'pholidotos' = scaly). There is a total of about 40 species in this genus, 8 of which are found in Malaya.

Pholidota gibbosa (Bl.) de Vr. [Figs 45, 46 & 47]
A highland epiphyte with closely spaced, longly ovate (prismatic) pseudobulbs, 6-8 cm long, 2 cm wide, two-leaved; greyish green with a dull surface. An interesting and attractive species. The leaves are lanceolate-acute, stalked, rather stiff; blade to 20 cm long, 5.2 cm wide; petiole to 4.5 cm long. Petioles and lower part of scape covered with large stem bracts, to 7 cm long. Inflorescence synanthous. Stalk very slender, 10 cm long, 1 mm wide. Rhachis pendulous from erect scape; slightly zig-zag, to 15 cm long, many-flowered. Flowers rather large, with conspicuous, transversely elliptical, brownish pink bracts, 1 cm long, 1.1 cm wide, falling when flower opens.
Sepals: Elliptical-acute, 8 mm long, 5 mm wide, very much reflexed, almost parallel with ovary.
Petals: Linear-truncate, 7 mm long, 0.6 mm wide, white, much reflexed, with incurved apex.
Lip: With cymbiform base and widened, much deflexed blade. Blade white, with two large, rounded lobes in front, and two small, triangular side lobes. At base of blade two curving, obliquely triangular calli. Base of lip pale mauve.
Column: Short, pale mauve, with very large wing. Column-wing tepal-like, widely obovate, 5.5 mm long, 6 mm wide; pale yellow, with brown margin in apical part.
Anther: Brown.
Distribution: Malaysia, Indonesia.

Pholidota parviflora Hook. f., syn. *Crinonia parviflora* Pfitz. [Figs 48 & 49]
A highland epiphyte with trailing rhizome and rather closely placed, cylindrical pseudobulbs, 2-3.5 cm long, 0.5 cm wide, two-leaved.
Leaves: Lanceolate-acute, stiff, 2-4.5 cm long, 1-1.5 cm wide.
Inflorescence: Synanthous, many-flowered, 4-6.5 cm long.
Bracts: Acute, pale mauve, 8 mm long, falling before flower opens.

Fig. 50. *Pholidota pallida* Lindl. August. 0.2×.

Fig. 51. *Pholidota pallida*. 3.5×.

Fig. 52. *Pholidota carnea* (Bl.) Lindl. Alt. 1650 m. January. 1.1×.

Fig. 53. *Pholidota carnea*. 5.4×.

Flowers: Start opening in apical part of inflorescence, although not at apex.
Sepals: Ovate-acute, 3-6 mm long, 2-3 mm wide, hyaline, white.
Petals: Ovate-acute, 3-5 mm long, 1.6-2.8 mm wide, hyaline, white.
Lip: Trilobed, white, with pale green hypochilium. Side lobes very indistinct, rounded, erect. Mid-lobe ovate-subacute with two keels; apex very sharply deflexed. Hypochilium saccate. Claw narrow, about 1 mm wide.
Column: Very pale mauve, with two porrect stelidia.
Anther: Brick-red.
Distribution: Endemic to Malaya.

Pholidota pallida Lindl. [Figs 50 & 51]
Epiphyte with sympodial growth. Pseudobulbs ovoid, with 3-4 indistinct ridges; subtruncate at apex; to 4 cm long, 2 cm wide; densely spaced; green; single-leaved.
Leaf: Oblanceolate-acute, to 32 cm long and 5.5 cm wide; slightly thickened; very tough, coriaceous; more or less plicate. Petiole 2-4 cm long.
Inflorescence: Synanthous, from top of young pseudobulb. Scape slender, arching, slightly flattened, 1 mm wide. Whole inflorescence to 38 cm long; rhachis 12-16 cm long, slightly zig-zag, many-flowered (18 to 40 flowers).
Bracts: Transversely elliptical-apiculate, much convex, distichous; 8 mm long, 10 mm wide; pale mauve and scurfy on open flowers; green, with mauve margins, before flower opens.
Pedicel + ovary: 5 mm long.
Dorsal sepal: Broadly ovate-obtuse, much convex; 7 mm long, 6 mm wide (when flattened); not much spreading.
Lateral sepals: Cymbiform-obtuse; 8 mm long, 3.5 mm wide (not flattened); with very sharp longitudinal wing on dorsal side.
Petals: Lanceolate-obtuse, slightly falcate; adnate to dorsal sepal; 6 mm long, 2 mm wide. All tepals pale mauve.
Lip: Trilobed, about 7 mm long. Mid-lobe about 3 mm long, deflexed; with two distinct lateral lobes. Side lobes erect – incurved over column; rounded. Base of lip saccate. Whole lip creamy white, thickened.
Column: 4 mm high, 3 mm wide (across wings); lateral and apical wings conspicuous, about 1.5 mm wide (or less). Operculum red-brick-coloured.
Distribution: From India and Sri Lanka through Thailand and Malaysia to Indonesia and New Guinea.

Pholidota carnea (Bl.) Lindl., syn. *Crinonia carnea* Bl. [Figs 52 & 53]
A highland epiphyte with stout, scaly rhizome, completely covered with large sheaths. Pseudobulbs obovoid, two-leaved, 1-2.5 cm long, 0.6-1.5 cm wide, pale green flushed with red; spaced 1-5 cm, and at 45° angle with rhizome.
Leaves: Lanceolate or oblong, acute, with red petiole 1-2 cm long; leaf-blade 5-15 cm long, 0.5-1.3 cm wide.
Inflorescence: Rather densely many-flowered, to 18 cm long; rhachis 15 cm long. Base of flowering stalk with large scales. Flowers start opening at apex of inflorescence.
Bracts: Fallen when flowers open.
Pedicel + ovary: 6 mm long.
Dorsal sepal: Ovate-rounded (or subacute), to 3.2 mm long, 2.2 mm wide, pink.

Fig. 54. *Chelonistele perakensis* (Rolfe) Ridl. 3.9×.

Fig. 55. *Coelogyne carnea* Hook. f. Alt. 1600 m. 2.4×.

Lateral sepals: Ovate-acute, to 3.5 mm long, 2.5 mm wide; mid-vein forming keel on dorsal side; pink.
Petals: Elliptical to obovate, apiculate (sometimes obtuse), 2.5 mm long, 1.5 mm wide, pink.
Lip: 3 mm long, indistinctly trilobed, pale yellow with salmon-coloured callus. Mid-lobe rounded and sharply deflexed at apex; side lobes small, erect.
Column: With two falcate, pointed arms, as long as anther; pink to brick-red, apex and column-foot darker.
Distribution: Thailand, Malaysia, Indonesia, New Guinea, Philippines.

GENUS CHELONISTELE Pfitzer, 1907

A small epiphytic genus, closely related to *Pholidota* and *Coelogyne*. The basal part of the lip is narrow, saccate, and not sharply distinct from the blade. The side lobes are narrow. There is probably only one species, but a variable one, in Malaya. The name of the genus seems to be derived from Gk 'kheilos' = a lip, and 'stele' = a post.

Chelonistele perakensis (Rolfe) Ridl., syn. *Chelonistele sulphurea* [Fig. 54]
A highland epiphyte with closely spaced, one-leaved, ovoid pseudobulbs, 3-5 cm long, 1.5 cm wide.
Leaf: Oblong-acute, blade 12-20 cm long, 3-4 cm wide; very tough and leathery; petiole 5-10 cm long.
Inflorescence: Erect, with many flowers (up to 16), about 6 flowers open at the same time. Scape entirely covered with three large, persistent, mauve-coloured sheaths.
Bracts: Ovate-acute, 12 mm long, 6 mm wide, mauve, normally all fallen when first flower opens.
Sepals: Lanceolate-acute, 15 mm long, 7 mm wide, spreading; mid-vein forming distinct keel on dorsal side; pale mauve or yellow.
Petals: Linear-acuminate, apex rolled backward; 10 mm long, 2 mm wide at base, pale yellow.
Lip: Trilobed, 15 mm long. Side-lobes tooth-like, 3.5 mm long. Mesochilium 4 mm wide, with two spreading keels. Epichilium widening to 9 mm, deeply cleft at apex, forming two rounded lobes. Whole lip white, with bright orange-yellow, irregular patch on epichilium, distal part of keels and on side lobes.
Column: 9 mm long, slightly curved, with distinct wings. Anther orange.
Distribution: Malaya, Sumatra, Java.

GENUS COELOGYNE Lindley, 1822

This is a large, South Asian genus with more than 150 species, many of which are very beautiful and of great horticultural value. The genus has a long history of cultivation, and a few hybrids have been produced. The pseudobulbs have one or two broad leaves which are often plicate. The inflorescence is pendulous or erect, with a few to many, medium-sized to rather large flowers. These sometimes open successively, in other species all the flowers are open at the same time. The lip is trilobed with erect side lobes close to the column, the mid-lobe being provided with a number of prominent longitudinal keels. The column is winged at the apex. Some 25 species have been

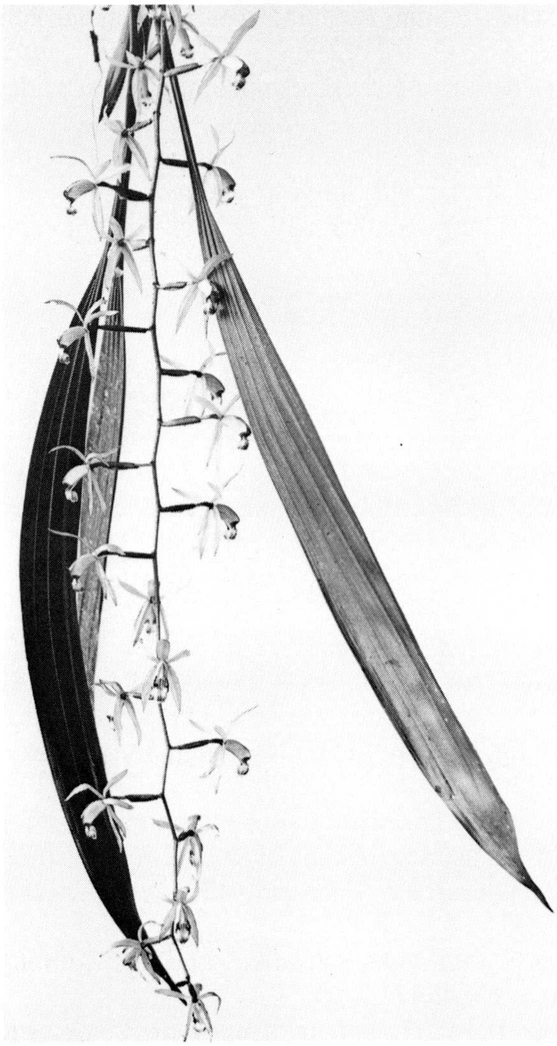

Fig. 56. *Coelogyne massangeana* Reichenb. f. Alt. 1600 m. July. 0.3×.

Fig. 57. *Coelogyne massangeana*. 0.9×.

recorded from Malaya. The origin of the generic name is uncertain: while some authors believe it to be derived from L. 'coelum' = heaven, + Gk 'gyne' = a woman, others prefer a more prosaic interpretation: from Gk 'koilos' = hollow + Gk 'gyne' = a woman, referring to the depression in the stigma.

Coelogyne carnea Hook. f. [Fig. 55]

A highland species, found in many places at altitudes above 1500 m. This is a variable species. Another common form has slightly smaller flowers with pale green tepals. The buds are light greenish brown, and the mesochilium has dark brown markings. The keels are long, thin and irregular. The pseudobulbs are more or less conical, and generally widening from the base upwards; they comprise one joint of the stem, of which a short, slender piece is normally seen below the pseudobulb, which is two-leaved, 2.5-8 cm long, and 0.8-2 cm wide.

Leaves: Petiolate; the narrowly ovate-acute blade is 4-20 cm long and 2-4 cm wide; the petiole is 0.5-3 cm long.

Inflorescence: Arises between the two leaves at the top of the pseudobulb; the scape is up to 10 cm long. The rhachis keeps growing as successive flowers develop, and can be found with a length of 30 cm, or more. It is zig-zag, with only one flower open at a time, at rather long intervals.

Bracts: Very large and generally dark mauve in colour. They fall before the flower opens.

Sepals: Lanceolate-acute, 20-25 mm long, 7-9 mm wide, hyaline, mauve.

Petals: Linear-acute, base slightly widened; 20 mm long, 1.5 mm wide (blade), hyaline, mauve.

Lip: 25 mm long, distinctly trilobed, deflexed. Side lobes erect, rounded, 4 mm high, 10 mm long, hyaline, mauve, with darker veins; mid-lobe spathulate. The sub-orbicular blade is 10 × 10 mm, and mauve, with indistinct venation on epichilium.

Column: Long and slender, winged around the top, mauve.

Coelogyne massangeana Reichenb. f. [Figs 56 & 57]

A large, attractive highland epiphyte which is often seen in cultivation. The pseudobulbs are up to 18 cm high and 3 cm wide, narrowly ovate, densely ribbed, slightly flattened and gradually tapering towards the apex. They have two leaves, and are often closely spaced on the rhizome. The leaves are narrowly obovate-acute, plicate, generally erect and up to 50 cm long, 9 cm wide (blade). The petiole is about 10 cm long. The inflorescence is limply pendulous, up to 80 cm long with many (about 25), large flowers. The rhachis is pale brownish mauve with dark brown hairs.

Bracts: The persistent bracts are much shorter than pedicel + ovary, and rolled-up around the pedicel; they are obovate-obtuse, 2.5 cm long, 1.8 cm wide, brownish mauve and slightly scurfy. The pedicels are 3.5 cm long, pale mauve with dark brown hairs. There are also dark brown hairs on the ovary and on the base of the dorsal side of the sepals.

Dorsal sepal: Lanceolate-acute, 36 mm long, 8 mm wide.

Lateral sepals: Obliquely lanceolate-acute, 35 mm long, 6 mm wide, pointing downwards.

Petals: Narrowly oblanceolate-acute, 32 mm long, 5 mm wide. All tepals are spreading, and pale mauvish yellow with mauve base.

Lip: Trilobed, 30 mm long, 21 mm wide across flattened side lobes. Side lobes erect, rounded (truncate in front), 8 mm long, 19 mm wide, cream-coloured; inner side

Fig. 58. *Coelogyne kaliana* Cribb. Pseudobulbs and leaves. 0.5×.

Fig. 59. *Coelogyne kaliana* Cribb.

Fig. 60. *Coelogyne kaliana*. Alt. 1650 m. September. 1.3×.

dark brown with white veins. Blade of lip with three parallel white keels which are fimbriate and increasing in height at base of lip (especially outer keels, which are also flushed with yellow on apices of longer, basal, fimbriate parts). Keels interspaced with brown lines, interrupted at base of mid-lobe, shortly before which keels turn yellow. Mid-lobe obcordate with point between frontal lobes; white, with brown venation. The point is slightly longer than the frontal lobes. The three keels are split up into thickened, warty cushion which is yellow, flushed with brown.

Column: Slightly arching, winged, 20 mm long, cream-coloured; ventral side brown. Operculum yellow. The flowers are rather unpleasantly scented (*Ligustrum vulgare*-like scent).

Distribution: Thailand, Malaya, Sumatra, Java.

Coelogyne kaliana Cribb [frontispiece & dust jacket; Figs 58, 59 & 60]

A beautiful highland epiphyte, recently discovered on the Main Range near Kuala Lumpur. It seems to prefer the highest peaks and rather exposed situations. The vegetative forms of this *Coelogyne* show much variation. The pseudobulbs range from a nearly globular form (the most common one in young individuals), typically 2.5 cm high and 2.8 cm wide, through ovoid shapes to a prismatic form, up to 5.5 cm long, 2 cm wide, with three distinct ridges, more often observed in old individuals. The pseudobulbs are always very closely placed, green, one-jointed and two-leaved. The leaves range from obovate to elliptical. They are always acute, with a blade up to 12 cm long, 5.5 cm wide; petiole up to 5 cm long. The leaves are not plicate, but stiff, thickened and leathery. The inflorescence is pendulous, up to 30 cm long, and 6-11-flowered. At the base, the scape is completely covered by stiff, scurfy, olive-coloured sheaths. All the flowers develop simultaneously. The rhachis is yellowish-brown, zig-zag, with large, obtuse to (sub)acute bracts with pale brown bases.

Bracts: Scurfy and persistent, covering the pedicel. The ovary is provided with short brown hairs.

Tepals: Lanceolate-acute, 45 mm long, 12 mm wide (petals slightly narrower than sepals); white.

Lip: Trilobed, 30 mm long. Side lobes rounded, erect, 10 mm long, 20 mm wide; outer side white, inner side dark brown with white venation. Mid-lobe 13 mm long, 11 mm wide, sub-orbicular with raised acute portion protruding at apex. Apex and margins of mid-lobe white, base dark brown with 6-7 yellow keels, two of which are provided with orange, fimbriate processes at base of lip.

Column: 20 mm long, widening from slender base, winged, white with brown striations in front, near base. Operculum cream-coloured. With rather unpleasant scent, resembling that of *Castanea sativa*-flowers.

GENUS EPIGENEIUM Gagnepain, 1932

A small genus with around 25 species, closely related to *Dendrobium*. Four species have been recorded for Malaya. The distribution is mainly South Asian.

The two-leaved pseudobulbs are conspicuously angled, and covered with large sheaths. Sepals and petals are similar, the lip is rather short. The genus can easily be confused with *Coelogyne*, but species of *Epigeneium* all have a mentum, as in genus *Dendrobium*. By some authors, *Epigeneium* is treated as a section of *Dendrobium*. The name of the genus is probably derived from Gk 'epi' = on + Gk 'geneion' = chin.

Fig. 61. *Epigeneium macropodum* (Hook. f.) Summerh. Alt. 1600 m. August. 0.5×.

Fig. 62. *Epigeneium macropodum*. 3.2×.

Fig. 63. *Epigeneium longipes* (Hook. f.) Summerh. Alt. 1650 m. January. 0.7×.

Fig. 64. *Epigeneium longipes*. 1.4×.

Epigeneium macropodum (Hook. f.) Summerh. [Figs 61 & 62]

A mountain epiphyte on high, exposed ridges. Pseudobulbs 3-5 cm long, 2.5 cm wide, two-leaved; well-spaced on sturdy (5 mm ∅) rhizome; obliquely ovoid-prismatic with four very sharp edges; surfaces between edges more or less plane. Pseudobulbs sometimes slightly flattened; pale yellowish green, flushed with purple, glossy; completely covered by large, thin sheaths (to 6 cm long).

Leaves: With short, 5-7 mm long, purple petiole. Blade elliptical, obtusely and almost equally bilobed at apex, 6-10 cm long, 2.3-3.1 cm wide, erect, stiff and leathery; upper side dark green, often with purple margins, glossy; with depressed mid-vein.

Inflorescence: From top of pseudobulb, between leaves; 3-12-flowered; slender deep purple stalk, with four acute, pale yellow, purple-veined stem bracts, 6 mm long. One larger sheath, 2.5 cm long, at base.

Floral bracts: Similar to stem bracts, 4-6 mm long, pale mauve.

Pedicel + ovary: Conspicuous, pale mauve to bright purple, 3-4 cm long, slightly curving.

Dorsal sepal: 16-19 mm long, 6-8 mm wide, lanceolate to oblong, acute, slightly recurved; pale yellow, with 6-7 mauve veins, darker on dorsal side, which is more or less flushed with purple.

Lateral sepals: Obliquely lanceolate-acute, 17-19 mm long, 8 mm wide, joined to column-foot; forming short (6-8 mm long) mentum which is perpendicular to ovary and lip; lateral sepals pale yellow with mauve veins; darker and more or less flushed with purple on dorsal side; spreading.

Petals: Lanceolate-acute, slightly oblique, 16-17 mm long, 5 mm wide, not much spreading; pale green (or yellow), with mauve veins which are darker on dorsal side.

Lip: Trilobed, 13 mm long. Side lobes erect, 7-8 mm wide, 3-4 mm high; apex slightly rounded, anterior edge truncate, pale green with irregular purple pattern, or wholly purple. Mid-lobe triangular, retuse at apex, 4.5 mm long, 6 mm wide; apex and margins white, centre lemon yellow with irregular purple streaks and numerous keels (about 7), most of which are short and indistinct. Central keel distinct, yellow, continuing almost to apex of mid-lobe.

Column: 10 mm long, much concave, with foot 5 mm long; yellowish green, with dense purple markings in front.

Anther: Yellow.

Distribution: Malaya, Sumatra.

Epigeneium longipes (Hook. f.) Summerh., syn. *Epigeneium kalianum* Stone [Figs 63 & 64]

A small highland epiphyte, growing in mist forest on exposed ridges. The pseudobulbs are rather closely placed (1 cm) on thick, branching rhizome; they are ovate, slightly flattened, with two sharp and two very blunt edges; 2 cm high, 1.5 cm wide, yellowish green, glossy when not covered with sheaths; two-leaved.

Leaves: Elliptic to sub-orbicular, rounded, with sharp incision at apex; 2 to 5 cm long, 2-4 cm wide, concave, stiff and leathery; with short petiole, to 0.5 cm long.

Inflorescence: From top of mature pseudobulb, slender (1 mm wide), erect, olive-reen, with three dark brown, tightly clasping sheaths. Scape 10-15 cm long, rhachis 1.5 cm long, with 2-5 rather large flowers (5 cm across).

Bracts: Very small, lanceolate-acute, light brown, 4 mm long.

Fig. 65. *Dendrobium laciniosum* Ridl. October. 1.3×.

Fig. 66. *Dendrobium laciniosum*. 8.6×.

Flowers: Pleasantly scented and long lasting (about two weeks). Flowering gregarious.
Pedicel + ovary: 2.5 cm long, very pale yellow.
Sepals: Lanceolate-acute, 25 mm long, 7 mm wide (laterals 8 mm wide), very pale yellow, spreading; widest in the middle.
Petals: Lanceolate-acute, tapering towards base and apex, 25 mm long, 7 mm wide, base 3 mm wide; very pale yellow, spreading.
Lip: Trilobed, 17 mm long. Mid-lobe triangular-acute, 8 mm long, 7 mm wide, slightly deflexed (esp. apex), very pale yellow; base of mid-lobe bright lemon-yellow. Lower side of lip with irregular, dark brown lines, continuing half-way up side lobes. Side lobes obliquely triangular-obtuse, erect, frontal margin perpendicular to blade; bright yellow; rest of side lobes white, with dark brown venation. Blade of lip with three interrupted keels (white) between side lobes.
Column: Slightly curved, 9 mm long, very pale yellow, winged. Column-foot bright orange-yellow, with brown band below stigma. Operculum cream-coloured.

GENUS DENDROBIUM Swartz, 1799

This is one of the largest genera of orchids: with about 900 species it is probably second in size to *Bulbophyllum* only. The distribution is South Asian-Australian, with representatives in New Zealand and Polynesia. Very few *Dendrobia* are terrestrial: most are epiphytic (Gk 'dendron' = a tree, 'bios' = life), including some very beautiful species, many of which are cultivated. A large number of hybrids have been produced. The wild Malayan species number about 100.

Vegetatively, there is considerable variation within the genus; the floral characteristics are, however, more constant. The lateral sepals have a wide base which is joined to the column-foot, the resulting structure being known as the mentum. The base of the lip is also joined to the column-foot. There are four pollinia, forming two pairs, without caudicles.

For convenience' sake, a large genus like *Dendrobium* is generally divided into sections. The arrangement below is essentially that of J.J. Smith, but with many sections left out.

Section Desmotrichum

Pseudobulbs of one node; widely spaced on branching rhizome and bearing a single leaf without sheaths. Flowers one or two, from tufts of dry bracts at base of leaf. Lip trilobed and provided with two or three keels on the blade. The flowers are short-lived (when the section is treated as a genus, it was previously known as *Ephemerantha*, and now as *Flickingeria* Hawkes, 1961).

Dendrobium laciniosum Ridl., syn. *Ephemerantha laciniosa* (Ridl.) P.F. Hunt & Summerh. [Figs 65 & 66]
A lowland epiphyte with narrowly cylindrical, flattened pdeudobulbs, 4.5 cm long, 0.7 cm wide. The single leaf is sessile, on top of the pseudobulb, and elliptical or lanceolate, acute, 8-12 cm long, 2.5 cm wide. The flowers arise among chaffy bracts, 5 mm long, at back of leaf. Pedicel + ovary 8 mm long.

Fig. 67. *Dendrobium kelsallii* Ridl. Alt. 1400 m. October. 0.8×.

Fig. 68. *Dendrobium kelsallii*. 3.7×.

Plate 2
2.1. *Dendrochilum carnosum* (Ridl.) Holttum. 5.0×.
2.2. *Dendrobium kelsallii* Ridl. 5.2×.
2.3. *Dendrobium hasseltii* (Bl.) Lindl. 3.5×.
2.4. *Dendrobium pahangense* Carr. 5.7×.
2.5. *Trichotosia microphylla* Bl. 8.3×.

2.1	2.2	2.3
2.4	2.5	

Dorsal sepal: Ovate-acute, 8 mm long, 3.5 mm wide, lowest 1.5 mm joined with laterals.

Lateral sepals: Obliquely triangular, 0.9 cm long, 0.7 cm wide at base, forming mentum. All sepals and pedicel + ovary pale yellow with minute brown spots (scurfy) on dorsal side.

Petals: Lanceolate-obtuse, with slight indentation in apical part, 8 mm long, 3 mm wide, pale yellow.

Lip: Deeply trilobed, 1.1 cm long. Mid-lobe with two large frontal lobes; the latter sub-orbicular, 3 mm Ø, up-curved, yellow. Blade yellow, with two wavy keels, convergent towards base, where they are parallel. Side lobes obliquely triangular, pale yellow, pointing forward and upward.

Column: Short, white, with two subulate processes pointing upwards.

Anther: Green.

Dendrobium kelsallii Ridl. [Plate 2.2; Figs 67 & 68]

A widely distributed mountain and lowland epiphyte with branched, elongated, trailing rhizome, 1.5 mm wide. Pseudobulbs one-leaved, spaced 2.5 cm, narrowly and obliquely fusiform with many fine ridges; glossy and often purplish; up to 3.5 cm long, 0.7 cm wide, forming narrow angle with continuing rhizome. Young pseudobulbs with smooth, glossy surface and tightly adpressed sheath, covering lower half of one side of pseudobulb.

Leaf: Lanceolate, very finely and (sub)acutely bilobed at apex; 4.5-7.5 cm long, 0.5-1.2 cm wide.

Inflorescence: Single-flowered, the flower arising among several small bracts behind leaf. Largest bract obtuse, 4 mm long, yellow.

Pedicel + ovary: Curving, flushed with purple, 8 mm long.

Dorsal sepal: Ovate-acute, 5-6 mm long, 3-4 mm wide.

Lateral sepals: Obliquely triangular-acute, 1.2 cm long from apex of sepal to apex of mentum. Mentum 7-11 mm long. All sepals pale yellowish mauve with purple veins; joined in basal 1 mm; spreading.

Petals: Linear-acute, with slightly widened base, 6 mm long, 1.5 mm wide, pale green or pale yellow with purple veins (or with pale purple flush).

Lip: Trilobed, 15 mm long. Side lobes rounded, erect, 2-3 mm long; very pale yellow, sometimes slightly flushed with purple. Blade with three conspicuous, parallel keels, slightly diverging at base of lip. Lateral keels widening and diverging near anterior end of side lobes. Middle keel 1 mm longer than lateral keels. Keels closely spaced, pale yellow with dark purple margins. Mid-lobe very deeply cleft, with rounded tooth near base, on outer margin; 5-6 mm long, 3.5-4 mm wide.

Column: 3 mm long, pale green, or white with pale green apex. Column-foot 6 mm long. Base of column yellowish green.

Anther: Pale green.

Distribution: Malaya, Sumatra.

Section Pedilonum

Stems fleshy, close together, with several internodes. No pseudobulbs. Leaves several, with sheaths. Inflorescences axillary, usually on leafless stems. Flowers two, or more, with a long mentum.

Fig. 69. *Dendrobium hasseltii* (Bl.) Lindl. Alt. 1650 m. June. 1.1×.

Fig. 70. *Dendrobium hasseltii*. 3.5×.

Fig. 71. *Dendrobium roseatum* Ridl. Alt. 1500 m. December. 1.3×.

Dendrobium hasseltii (Bl.) Lindl. [Plate 2.3; Figs 69 & 70]

A common highland epiphyte with slender stems, thickened in upper third, 20-60 cm long, with leaves in upper half.

Leaves: Lanceolate, unequally and acutely bilobed at apex, 3-10 cm long, 1-1.5 cm wide.

Inflorescences: 1-5, near apices of leafless stems; 2-6-flowered.

Bracts: Ovate-acute, 5 mm long, mauve.

Flowers: To 3.2 cm long from apex of dorsal sepal to apex of mentum; pendulous, not resupinate, persistent.

Pedicel + ovary: 2.5-3 cm long, deep purple, sometimes slightly S-shaped.

Dorsal sepal: Ovate-subacute, 9-13 mm long, 6-7 mm wide, purple.

Lateral sepals: Ovate-acute, joined at base which is drawn out to very long mentum. Length of lateral sepal from apex to end of mentum: 25 mm. Length of mentum from onset of ovary: 16-19 mm.

Petals: Ovate-acute, 9-12 mm long, 5-7 mm wide, purple.

Lip: Spathulate-subacute, 11 mm long, 5-6 mm wide (blade); pale purple with large yellow-orange spot on blade.

Column: White or pale purple.

Anther: White.

Dendrobium roseatum Ridl. [Fig. 71]

Highland epiphyte with long, erect, slender, tufted stems, which are widest above the middle, to 60 cm long; 2 mm wide at base, 7-8 mm wide in the middle of stem (which is not flattened). Internodes 2.5-3 cm long.

Leaves: Lanceolate-acute, 7-14 cm long, 1.2-2 cm wide, present in upper part of stem only.

Inflorescences: Only on leafless stems, 2-5-flowered.

Bracts: Lanceolate-acute, 3.5 mm long, green.

Pedicel + ovary: To 2.5 cm long, pale mauve.

Flowers: White, persistent (lasting more than a week).

Dorsal sepal: Ovate-rounded, 12-14 mm long, 7-8 mm wide, dorsal side with very small apiculus near apex of dorsal side; white, with very slight lilac flush near base.

Lateral sepals: Blade ovate-rounded, 12 mm long, with apiculus similar to the one on dorsal sepal; base drawn out to long mentum. Dorsal side of sepals, including mentum, with slight lilac flush. Mentum 19 mm long, widening toward apex, but laterally flattened in apical 3 mm. Sac-like part 4.5 mm wide, narrowest 3 mm wide; sepals joined for 10 mm, at apical part of mentum.

Petals: Obovate-rounded, with wavy margins; 12-14 mm long, 9 mm wide; sometimes with slight lilac flush on median line of dorsal side.

Lip: Very stiff, attached to column-foot, 13 mm long, 2.5 mm wide; margins up-turned; base very slender. Blade much widened, nearly square, slightly reflexed above middle, with wavy margins, 11 × 11 mm; pure white, with central groove starting at deep incision at apex (not clearly visible due to undulating margin).

Column: Short (4 mm high); with rounded, pale yellow, lateral wings between anther and stigma (about as high as anther), with two short, dark yellow, tooth-like projections near stigma. Column-foot consisting of two long, white keels, slightly diverging towards point of attachment of lip; 16 mm long. Operculum white.

Distribution: Endemic.

Fig. 72. *Dendrobium bifarium* Lindl. ex Hook. f. May. 1.3×.

Fig. 73. *Dendrobium bifarium*. 6×.

Fig. 74. *Dendrobium pahangense* Carr. Alt. 1200 m. July. 3.7×.

Fig. 75. *Dendrobium pahangense*. 5.2×.

Section Distichophyllum

Stems close together, rather slender, leafy, with short internodes. Leaves small, rounded, two-ranked. Flowers 1-3, on the leafy stems. Lip straight.

Dendrobium bifarium Lindl. ex Hook. f. [Figs 72 & 73]

Lowland epiphyte with many, densely crowded stems, more or less erect, to 50 cm high, leafy only in upper part, not much thickened (to 8 mm wide). Internodes 6-10 mm long.

Leaves: Thin, with tightly clasping sheaths; narrowly elliptical, unequally and obtusely bilobed at apex, 2-5 cm long, 1 cm wide.

Inflorescences: Lateral, single-flowered.

Bracts: Minute, triangular-acute, 1.5 mm long.

Dorsal sepal: Lanceolate-acute, much reflexed, 5-7 mm long, 2.5-3 mm wide, white, with pale green apex.

Lateral sepals: Oblong-acute, bases forming mentum. Laterals 11 mm long from apex to end of mentum, white, with pale green apex. Mentum long, slender, parallel with ovary; 5 mm long.

Petals: Narrowly lanceolate-acute, 8 mm long, 1.2-2 mm wide, white.

Lip: 13 mm long, with much widened blade from narrow base; blade 7 mm wide. Base white, blade green; centre of blade papillose, pale green to white or yellow; with three green keels.

Column: Wide, white.

Anther: Creamy white.

Distribution: Thailand, Malaysia.

Dendrobium pahangense Carr [Plate 2.4; Figs 74 & 75]

Highland epiphyte with slightly thickened (4 mm wide), closely placed stems, internodes 4 mm long. Stems up to 30 cm long, leafy at apex.

Leaves: Two-ranked, with imbricate, yellowish-green sheaths; elliptic, obtusely bilobed at apex, up to 20 mm long, 6 mm wide.

Inflorescences: From leaf-axils, single-flowered.

Bracts: 2 mm long, ovate-acute.

Pedicel + ovary: 8 mm long.

Dorsal sepal: Lanceolate-acute, 6-7 mm long, with recurved margins.

Laterals: Upper part triangular-acute, 7 mm long, 4 mm wide; lower part forming long, spur-like mentum, 6-7 mm long from onset of ovary.

Petals: Lanceolate-acute, 6-7 mm long, 1.5-2 mm wide. All tepals white.

Lip: White, trilobed, 13-15 mm long, with indistinct side lobes. Mid-lobe triangular-obtuse with wavy margin, decurved, 7 mm wide, 6 mm long. Blade of lip and centre of mid-lobe densely papillose, olive-green with three darker longitudinal streaks, changing to brown on hypochilium. Side lobes rounded, spreading, white, 0.5 mm long.

Column: White, with brown base.

Anther: Olive green.

Distribution: Probably endemic.

Section Rhopalanthe

Stem slender except for swollen part (one or a few internodes) near the base. Base and

Fig. 76. *Dendrobium crumenatum* Sw. October. 1.3×.

Fig. 77. *Dendrobium leonis* (Lindl.) Reichenb. f. January. 2.7×.

Fig. 78. *Dendrobium leonis*. 4.5×.

apex of stem leafless. Flowers axillary from tufts of small dry bracts on leafless, upper part of stem, blooms very briefly.

Dendrobium crumenatum Sw. [Fig. 76]

A very common lowland epiphyte, known as the 'Pigeon Orchid' (since the buds look somewhat like little white pigeons). Pseudobulbs at 1.5-2 cm from base of stem, narrowly ovate, not flattened, or very slightly flattened, ridges often not very conspicuous; to 10 cm long, 2 cm wide, sometimes purple. Stems 30 cm long, or more, sometimes purple, covered with scarious sheaths, 2-3 cm long.

Leaves: Jointed at upper edge of sheaths, lanceolate to elliptical, obtusely and more or less equally bilobed at apex; to 6 cm long, 2 cm wide, yellowish green (sometimes with purple spots), rather thick and stiff.

Inflorescence: Many-flowered, on leafless, apical part of stem; rhachis 10 cm long, or more. Flowering gregarious.

Bracts: Scarious, 9 mm long.

Flowers: White, with very strong fragrance, but not long lasting (less than one day).

Pedicel + ovary: 12 mm long; greenish white, yellow near sepals.

Dorsal sepal: Lanceolate-acute, slightly widened 3 mm from apex; 20 mm long, 6.5 mm wide, spreading.

Lateral sepals: Obliquely lanceolate-acute, spreading; base drawn out to form long, spur-like mentum. Length of lateral sepal from onset of ovary 15 mm. Distal half of mentum deflexed, more or less parallel with rhachis. Apex of mentum with constriction in last 1 mm; flushed with yellow. Width of lateral sepals 9 mm.

Petals: Ovate-acute, spreading, 18 mm long, 7 mm wide.

Lip: Trilobed, 25 mm long, 13 mm wide (across side lobes). Mid-lobe more or less transversely elliptical with fine point at apex; wavy margins, deeply lacerate near side lobes; three white keels, converging at apex. Mid-lobe 9 mm long, 11 mm wide. Keels turn bright yellow and increase in number to 6 between mid- and side lobes, and change again to pale yellow in basal half of lip. Side lobes obliquely triangular with wide base, 4 mm high, 11 mm long, erect. Whole lip decurved.

Column: Short, white, with long foot and two upwards-curving, tooth-like processes.

Anther: White.

Distribution: India (probably introduced), Sri Lanka (common, but probably introduced; often seen in cultivation), Andaman Islands, Burma, Thailand, Laos, Cambodia, Vietnam, China, Taiwan, Malaysia, Indonesia, Christmas Isl., Philippines.

Section Aporum

Stems and leaves laterally flattened. Leaves broad, thick and fleshy.

Dendrobium leonis (Lindl.) Reichenb. f. [Figs 77 & 78]

A lowland epiphyte with numerous, tufted stems, to 25 cm long, leafy throughout or in apical part only. Leaves much thickened and laterally flattened (no groove visible on upper edge); very stiff, closely spaced. Lower edge bluntly triangular to nearly straight, upper edge curved; with very tightly clasping and thickened sheaths.

Leaves: 14-25 mm long (including sheath), 7-12 mm high.

Inflorescence: Terminal, single-flowered, from tuft of scarious bracts.

Pedicel + ovary: 5 mm long. Flowers fragrant (*Vanilla*-like scent).

Fig. 79. *Dendrobium aloifolium* (Bl.) Reichenb. f. November. 11.6×.

Fig. 80. *Dendrobium sinuatum* (Lindl.) Lindl. Alt. 1400 m. July. 2.2×.

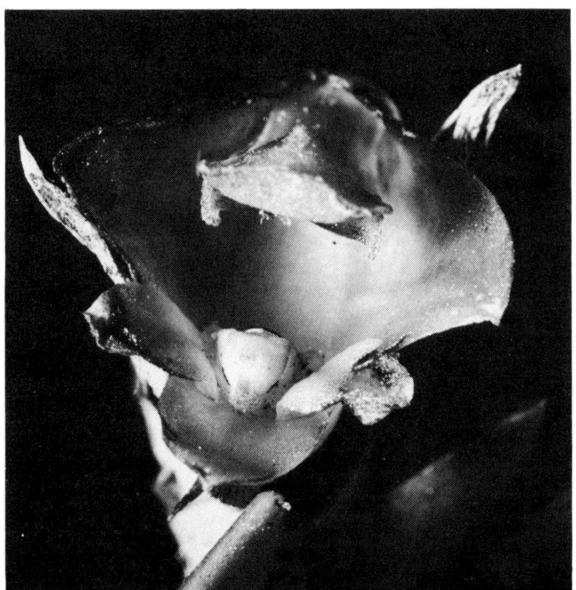

Fig. 81. *Dendrobium sinuatum*. 7.1×.

Dorsal sepal: Ovate-rounded, 6-8 mm long, 5 mm wide.
Lateral sepals: Obliquely triangular-obtuse, with large base, forming mentum, 10-12 mm long from ovary; spreading, with recurved apex (applies to all sepals). All tepals pale green or pale yellow, thickened, flushed with purple on dorsal side.
Petals: Rhomboid or narrowly elliptic, obtuse; with narrow base; 5.5-7 mm long, 3 mm wide, spreading.
Lip: Oblong, unlobed, 10-14 mm long, 5 mm wide, decurved; very pale green, flushed with purple (darker than sepals, esp. lower side of lip). Apex with two rounded lobes; margins white, fimbriate. Low, rounded, papillose, pale green callus between frontal lobes; often a purple spot below callus. Yellow to orange keel in basal part of lip, widening toward column-foot.
Column: Short, pale green, with one short, bluntly triangular point on each side of anther. Column-foot with purple streaks near apex.
Anther: Pale green.
Distribution: Laos, Cambodia, Vietnam, Thailand, Malaysia, Sumatra.

Dendrobium aloifolium (Bl.) Reichenb. f. [Fig. 79]
Lowland epiphyte, also found as a lithophyte on rocks near the sea. Stems elongated, to 60 cm long, much flattened; branching at base only, or rarely with lateral branches and adventitious roots.
Leaves: Thick, laterally much flattened (no upper surface or groove); broadly lanceolate-acute, with large sheath, clasping stem; to 30 mm long, including sheath; 7-10 mm wide; stiff, alternate, with 2-3 mm spacing between successive leaves on one side; internodes 80 mm long.
Inflorescence: Apical, with one flower open at a time, often at apex of continuously growing rhachis (to 32 cm long).
Bracts: Large bracts on rhachis lanceolate-acute, 11 mm long, laterally compressed.
Flowers: Very small, pale green, sometimes very slightly flushed with purple; from tuft of scurfy bracts.
Pedicel + ovary: 3.5 mm long.
Sepals and petals: Much spreading or reflexed.
Dorsal sepal: Ovate-subacute, to 2 mm long, 1-1.5 mm wide.
Lateral sepals: Ovate-subacute, 1.5 mm wide; with wide base, forming mentum, 2 mm long from onset of ovary.
Petals: Linear-acute, 2 mm long, 0.5 mm wide.
Lip: 3.5-4 mm long, trilobed, erect, in contact with column. Side lobes small, rounded erect. Mid-lobe spathulate; deeply cleft in middle, forming two rounded lobes.
Column: Short, pale green. Operculum green.
Distribution: Burma, Laos, Vietnam, Thailand, Malaysia, Indonesia.

Section Oxystophyllum

Leaves much flattened, narrower than in section *Aporum*; more imbricate at base. Stems always leafy throughout.

Dendrobium sinuatum (Lindl.) Lindl. [Figs 80 & 81]
Highland epiphyte with several thin, elongated stems, up to 60 cm long, but often much shorter, leafy throughout.
Leaves: Numerous, laterally flattened, imbricate at base, not incurved; having more or

Fig. 82. *Dendrobium carnosum* (Blume) Reichenb. f. April. 1×.

Fig. 83. *Dendrobium carnosum*. 3.8×.

Fig. 84. *Dendrobium excavatum* (Bl.) Miq. Alt. 1050 m. December. 1.8×.

Fig. 85. *Dendrobium excavatum*. 9.7×.

less the same length throughout, although slightly longer in middle of stem in young individuals; thick and stiff with sharp points; up to 40 mm long, 8 mm wide; rather dark green.

Inflorescences: From leaf axils, among tuft of chaffy bracts; many-flowered, but flowering singly in succession, with long intervals.

Bracts: Triangular-acute (or acuminate), 10 mm long, 4 mm wide.

Dorsal sepal: Narrowly elliptic-apiculate, 9 mm long, 3 mm wide.

Lateral sepals: Obliquely triangular-apiculate, 9 mm long.

Petals: Obovate-apiculate, sometimes with a few teeth at apex; 6 mm long, 2 mm wide. All tepals pale yellow or pale green, flushed with purple.

Lip: Uppermost, 9 mm long, very indistinctly trilobed. Mid-lobe thickened, with a point on lower side; transversely elliptic-obtuse, 4 mm wide, with irregular margin and very small notch at apex. Side lobes less than 1 mm long, with irregular margins. Lip mauve or purple, mid-lobe with very diffuse, pale yellow or whitish spot.

Column: Pale yellow, base of foot with purple streaks.

Anther: Yellow.

Distribution: Probably endemic to Malaya.

Dendrobium carnosum (Bl.) Reichenb. f., syn. *Dendrobium concinnum* Miq. [Figs 82 & 83]

A lowland epiphyte. Stems closely placed, to 30 cm long, slightly curving.

Leaves: 40-50 mm long, 5 mm wide, laterally compressed, closely imbricate at base, slightly incurved, lanceolate-acute, yellowish green.

Inflorescence: Many-flowered, but flowers appear singly, or in pairs. Flowers 8 mm long, not resupinate.

Bracts: Numerous, imbricate, very acute, laterally compressed, with raised, flattened mid-vein. Upper bracts acuminate.

Dorsal sepal: Oblong-apiculate, 5 mm long.

Lateral sepals: 6 mm long, obliquely triangular-apiculate, forming mentum, 4 mm long.

Petals: Linear-apiculate, 3-4 mm long, 1 mm wide.

Lip: Ligulate-apiculate, retuse, with apiculus on lower side of lip, about 1 mm from apex; apex slightly widening and swollen; fimbriate. Side lobes very small, erect, rounded.

Tepals: All pale greenish yellow with purple spots, denser and darker on blade of lip and column-foot. Outer side of tepals often without spots, especially apex of lip.

Distribution: Burma, Cambodia, Vietnam, Thailand, Malaysia, Indonesia.

Dendrobium excavatum (Bl.) Miq. [Figs 84 & 85]

A small mountain epiphyte, similar to *Dendrobium sinuatum*, but smaller in all parts. Stems numerous, densely tufted, about 5 cm long, leafy in upper part.

Leaves: Thickened, laterally compressed, closely two-ranked, lanceolate-acute, to 30 mm long, 4-5 mm wide, closely imbricate at base, slightly incurved.

Inflorescences: Apparently terminal, from tuft of chaffy bracts. Flowers not resupinate; purple in all parts except pale yellow apiculi. Tepals not much spreading.

Dorsal sepal: Lanceolate-acute, 4 mm long, 1.2 mm wide.

Lateral sepals: Obliquely triangular-acuminate, 4.5 mm long; with wide base, forming mentum, 2 mm long.

Fig. 86. *Dendrobium subulatum* (Bl.) Lindl. Alt. 550 m. November. 0.8×.

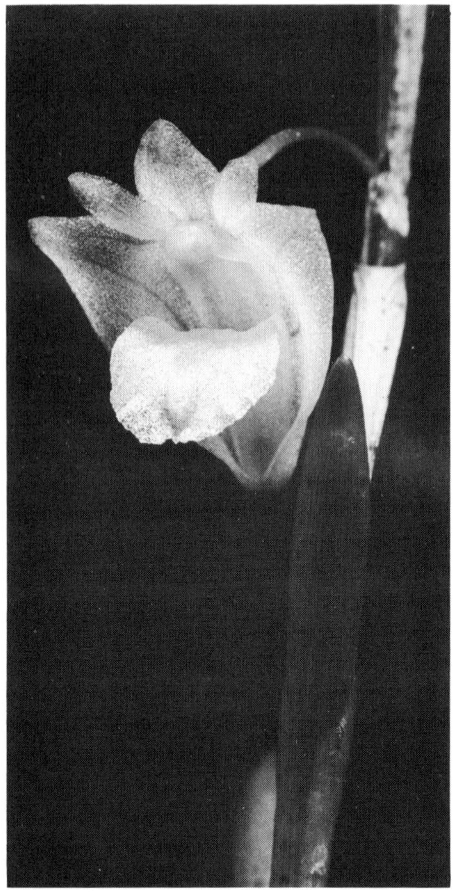

Fig. 87. *Dendrobium subulatum*. 7.1×.

Fig. 88. *Dendrobium kentrophyllum* Hook. f. Alt. 1050 m. December. 2.7×.

Fig. 89. *Dendrobium kentrophyllum* 5.1×.

Petals: Lanceolate-acute, 3 mm long.

Lip: Oblong, with rounded apex; 4 mm long, 1.5 mm wide. Apex with fimbriate margin and reniform, very dark purple, transverse callus; lower side of apex with rather sharp point. Blade with slightly raised, curving margins. Base of lip dark purple.

Column: Pale purple.

Anther: Purple.

Distribution: Cambodia, Malaysia, Indonesia, New Guinea.

Section Strongyle

Stems thin, leafy in lower part. Leaves laterally flattened or terete, not imbricate at base. Flowers from upper, leafless part of stem.

Dendrobium subulatum (Bl.) Lindl. [Figs 86 & 87]

Small lowland epiphyte. Stems to 20 cm long, rather densely tufted, leafy in lower part. Leafy part to 8 cm long, leafless part to 12 cm, covered with withered sheaths; internodes about 1 cm long, with very narrow base (0.5 mm wide), each internode widening to about 1.5 mm in upper part.

Leaves: About six, fusiform-terete, acute, at 45° angle with stem, arching; to 40 mm long (often shorter), 2.5 mm wide. Leaf-sheaths about 10 mm long, green, completely covering stem.

Inflorescences: From leafless part of stem, axillary, flowers arising singly among chaffy bracts.

Pedicel + ovary: 7 mm long; mauve, rarely pale green. Flowers 9 mm long from apex of dorsal sepal to apex of mentum.

Dorsal sepal: Triangular, obtuse to acute, 3 mm long, 1.5 mm wide, white with mauve base.

Lateral sepals: Triangular, obtuse to acute, with very wide base, forming up-curved mentum, 6 mm long; white, apex of mentum pale yellow.

Petals: Oblong to oblanceolate, rounded to acute, 2.5 mm long, 0.5 mm wide; white, with pale mauve base; sometimes pale pink, with pink veins.

Lip: Decurved, 8 mm long, 3.5 mm wide, unlobed, widening towards apex. Apex with more or less wavy margin and deep incision in front (or entire, with apiculus); white, with pale green keel on blade, darkest near apex.

Column: Short, about 1 mm; white, flushed with purple. Column-foot 5 mm long, white, with indistinct, 1.5 mm long, yellowish-green keel near claw.

Anther: White.

Distribution: Thailand, Malaysia, Indonesia.

Dendrobium kentrophyllum Hook. f. [Figs 88 & 89]

Highland epiphyte. Stem tufted, 15-20 cm long, entirely covered with leaf-sheaths; internodes about 8 mm long.

Leaves: Flattened, two-ranked, much thickened, terete-acute, 25-35 mm long, 4-6 mm wide.

Inflorescence: With one flower at a time, apical, from tuft of chaffy bracts.

Pedicel + ovary: 14 mm long.

Dorsal sepal: Lanceolate-subacute, 7-8 mm long, 3-3.5 mm wide, not much spreading.

Fig. 90. *Trichotosia ferox* Bl.

Fig. 91. *Trichotosia microphylla* Bl. Alt. 1060 m. October. 2.0×.

Fig. 92. *Trichotosia microphylla*. 6.7×.

Lateral sepals: Triangular-obtuse from wide base; 8 mm long. Mentum curving forward, 10 mm long. All sepals pale green, with pale purple veins.
Petals: Oblanceolate-rounded, 6 mm long, 1.8 mm wide, pale green.
Lip: Unlobed, 15 mm long; pale green, with wide central keel, and papillose, green callus near apex. Blade with raised, rounded margins. Apex rounded, wavy, with deep incision in front.
Column: Pale green, with purple foot.
Anther: White.
Distribution: Thailand, Malaya, Sumatra.

GENUS TRICHOTOSIA Blume, 1825

Closely related to *Eria*, and often included in this genus. All species have hairy stems and leaf-sheaths, and sometimes hairy leaves as well (Gk 'thrix' = hair). The hairs are usually stiff and reddish-brown. The dorsal side of the sepals is red-hairy. In Malaya, there are some 12 species.

Trichotosia ferox Bl., syn. *Eria ferox* (Bl.) Bl. [Fig. 90]
Large, sturdy highland epiphyte, with erect, very stiff stems; to 2 m high, branching only at base; 6 mm wide.
Leaves: Elliptic-acuminate, with long sheath, completely covering stem; 10 cm long, 3.5 cm wide. Whole leaf-surface, including sheath, with long hairs (1.5 mm long on blade, 3 mm long on sheath). Base of leaf twisted, blades more or less in one plane. Rhachis about 10 cm long, with short, bristly, reddish-brown hairs.
Inflorescences: Axillary, flowers about 10.
Bracts: Broadly triangular-acuminate, 9 mm long, 7 mm wide, with short, reddish-brown hairs.
Pedicel + ovary: 7 mm long, with dense, reddish-brown hairs. With strong, honey-like fragrance.
Dorsal sepal: Lanceolate-obtuse, 7 mm long, 4 mm wide; dorsal side with sparse reddish-brown hairs.
Lateral sepals: Broadly lanceolate-obtuse, falcate, 6 mm long (excl. mentum), 6 mm wide; sparsely hairy on dorsal side; mentum 7 mm long.
Petals: Oblanceolate-obtuse, apices spreading; 9 mm long, 1 mm wide at base, blade 3 mm wide. All tepals yellow-green, slightly thickened.
Lip: Trilobed, much deflexed, 14 mm long. Side lobes erect, with incurved edges and irregular teeth; 8 mm wide, 2 mm high; yellow, with faint purple markings. Mid-lobe transversely elliptical, apex sharply deflexed; edges sharply incurved; 6 mm wide, 4 mm long; yellow, flushed with purple; edges with irregular teeth. Blade of mid-lobe with irregular calli, and very short, dark purple hairs. Base of lip with long, low, narrow (1 mm) keel, with short white and purple hairs.
Column: 12 mm long, widening towards apex, curved; white, with pale brown hairs (or pale purple hairs). Operculum white.
Distribution: Thailand, Malaysia, Indonesia.

Trichotosia microphylla Bl., syn. *Eria microphylla* (Bl.) Bl. [Plate 2.5; Figs 91 & 92]
A highland epiphyte. Rhizome long, creeping, with numerous roots. Leafy shoots to 10 cm long, many-leaved, mainly in upper part. Rhizome and erect stems hairy.

Fig. 93. *Trichotosia gracilis* (Hook. f.) Kränzl. September. 1.5×.

Fig. 94. *Trichotosia gracilis*. 5.7×.

Leaves: Two-ranked, elliptical-rounded, rather thick, to 13 mm long, 7 mm wide, densely hairy; hairs white or brown, longer on lower side.

Inflorescences: 1-2-flowered, with dense long hairs; pedicel very short. Flowers persistent.

Bracts: Of two kinds – subulate and ovate-acuminate, present at the same time; 4-6 mm long.

Sepals: Densely hairy on dorsal side, ventral side glabrous; pale yellow.

Dorsal sepal: Lanceolate-acute, 6-7 mm long, 3-4 mm wide, slightly decurved.

Lateral sepals: Obliquely triangular-obtuse, 7-10 mm long, with wide base, forming mentum. Mentum 8 mm long, from onset of ovary.

Petals: Oblong-rounded or obtuse, glabrous, 5-7 mm long, 2 mm wide, pale yellow.

Lip: Ligulate, rounded (nearly truncate); 10 mm long, 4-5 mm across widened part of base, 3-4 mm across blade; apex swollen, papillose; base hairy on outer side. Small conical projection on lower side of blade. Margins of blade very slightly raised. Base of lip in contact with column; apical half deflexed, more or less parallel with column, touching apices of petals.

Column: Slender, pale green or yellow, with along foot; apex covered by lip. Anther off-white.

Capsule: Ovate, rounded, 7 × 5 mm; very hairy, resembling a small gooseberry.

Distribution: Thailand, Vietnam, Malaysia, Indonesia.

Trichotosia gracilis (Hook. f.) Kränzl., syn. *Eria gracilis* Hook. f. [Figs 93 & 94]

A lowland epiphyte with numerous, closely placed, erect to sub-erect, slightly hairy stems, to 20 cm long. Stems rather thin, internodes 1-1.5 cm long.

Leaves: Lanceolate-acute, very thick, to 6 cm long, 1.3 cm wide, with clasping sheaths. Mid-vein depressed. Leaves predominantly two-ranked.

Inflorescences: Numerous, axillary, (1)-2-flowered. Flowers long persistent (two weeks); with faint fragrance.

Bracts: Several, covering pedicel and ovary; upper bract triangular-acute, 8-12 mm long, 6-9 mm wide, very stiff, covering ovary and base of dorsal sepal. All bracts with dense, reddish-brown hairs.

Dorsal sepal: Oblanceolate-acute, 9-11 mm long, 3-5 mm wide, very pale mauve.

Laterals: Obliquely ovate-acute, apex conduplicate; 13-15 mm long from apex of mentum; 5 mm wide (blade), very pale mauve with darker mentum; apex pale green. Mentum saccate, up-turned apex touching ovary; 4-5 mm long. Dorsal side of sepals with brown hairs; red hairs on mentum.

Petals: Narrowly oblanceolate-rounded (or obtuse), 8.5-10 mm long, 2-2.5 mm wide, white.

Lip: Unlobed, cuneate-retuse, apex with depressed margins; 11-13 mm long, 4 mm wide; white, with dark purple margins on mesochilium and hypochilium. Base of lip purple. Blade in contact with column, pale yellow, with orange-yellow, triangular, hairy keel and two large patches of bright orange-yellow hairs.

Column: Wide (3.5 mm), white, with purple streaks in basal part. Operculum very dark purple, almost black. Ovary with purple lines.

Distribution: Thailand, Malaysia.

GENUS ERIA Lindley, 1825

This epiphytic genus is closely related to *Dendrobium*: the vegetative parts and the

Fig. 95. *Eria pulchella* Lindl.
Alt. 1050 m. November.
4.3×.

Fig. 96. *Eria leiophylla* Lindl. October. 0.4×.

Fig. 97. *Eria leiophylla*. 1.6×.

flowers are similar in structure. *Eria* is, however, distinguished by the following features: base of lip not joined to column-foot; pollinia 8, in two groups of 4, with caudicles.

Eria is a large genus, with about 400 species, of which some 60 have been recorded for Malaya. Erias are rarely cultivated since the flowers in most species are rather small. However some of the small-flowered species, such as *Eria pachystachya* (page 79), can be quite attractive when bearing a number of dense, many-flowered, elongated inflorescences.

Eria pulchella (see below) is a beauty by any standard: the flowers are comparatively large, with pale orange, white-hairy sepals and petals, the lip is very dark brown and the column black velvet! The name of the genus comes from the Greek 'erion', meaning 'wool' – indeed, many Erias are conspicuously woolly.

Eria pulchella Lindl. [Plate 3.1; Fig. 95]

Highland epiphyte with elongated, winding and branching rhizome. Pseudobulbs 4-angled, flattened, narrowly ovoid, 5-6 cm long, 2 cm wide, 1.3 cm thick; stem below pseudobulb 1.5 cm long. Pseudobulbs olive-green, without sheaths. Leaves three; first leaf at apex of pseudobulb; the other two leaves on successive swollen joints of stem (much smaller than pseudobulb).

Leaves: Oblong-rounded and slightly bilobed at apex; 10-12 cm long, 2.8 cm wide (lowest leaf; upper leaves slightly smaller).

Inflorescences: To 8 cm long, axillary from bulbous swellings, many-flowered, only one flower open at a time. Inflorescences, bracts, pedicels, ovaries and sepals covered with dense, short white hairs.

Bracts: Very broadly ovate to circular, rounded at apex, 4 × 4 mm.

Flowers: Last two days, turning orange when old.

Pedicel + ovary: 2 cm long.

Sepals: Thickened, pale yellow tinged with mauve; dense white hairs even on ventral side. All tepals much spreading.

Dorsal sepal: Lanceolate-acute, 11-15 mm long, 5-6 mm wide.

Lateral sepals: Obliquely lanceolate-acute, 10-14 mm long, 5-6 mm wide, mid-vein forming keel on dorsal side.

Petals: Oblanceolate, rounded or obtuse, 9-10 mm long, 3 mm wide, ventral side with dense white hairs, dorsal side almost glabrous.

Lip: Attached to column-foot with narrow claw; cordate-obtuse; 6 mm long, 4.5 mm wide, convex; covered with fine brown hairs. Blade with wide keel, 3.5 mm long, 2 mm wide, not much raised, dark brown at apex, base lighter brown, glabrous.

Column: Curved, 6 mm long, yellow. Column-foot 3.5 mm long, 2 mm wide, with numerous calli arranged fanwise. Column-foot and ventral side of column conspicuously deep black, with some very fine, white hairs. Anther greenish-yellow.

Distribution: Thailand, Malaysia.

Eria leiophylla Lindl. [Plate 3.2; Figs 96 & 97]

Lowland epiphyte. Pseudobulbs on stout rhizome; ovate, slightly flattened, 3 cm long, 2 cm wide. Stem above pseudobulb tapering, with a number of brown-hairy sheaths and two leaves at apex.

Leaves: Oblong; obtusely, unequally and conniventLy bilobed at apex (not very conspicuously); to 17 cm long, 2.2-2.5 cm wide, slightly arching in apical half. Base of leaves sparsely brown-hairy.

Fig. 98. *Eria longifolia* Hook. f. Alt. 1400 m. July. 0.6×.

Fig. 99. *Eria longifolia*. 6.7×.

Fig. 100. *Eria robusta* (Bl.) Lindl. Alt. 1600 m. June. 0.5×.

Inflorescences: Lateral, short-stemmed (to 1 cm long), covered with dense reddish-brown hairs; two-flowered.

Dorsal sepal: Ovate-apiculate, 13-15 mm long, 6 mm wide.

Laterals: Obliquely triangular-apiculate, 15 mm long, 9 mm wide; forming mentum, 7 mm long. Dorsal side of sepals densely covered with matted brown hairs.

Petals: Lanceolate-rounded, 11 mm long, 4.5 mm wide. All tepals orange-yellow, flushed with purple especially in centre; with fine, white hairs on ventral side.

Lip: 13 mm long, 6 mm wide (blade without side lobes); trilobed, slightly arching. Mid-lobe ligulate, much swollen, papillose, very dark purple. Large irregular, papillose keel throughout length of lip. Keels forms arrow-like structure with apical swollen part. Mid-lobe 5 mm long, 4.5 mm wide. Side lobes 3 mm high; rounded, erect, purple. Blade of lip purple.

Column: Arching, pale yellow; end of column-foot purple; column with brownish-purple line below rostellum.

Anther: Yellow.

Distribution: Malaya, Indonesia.

Eria longifolia Hook. f., syn. *Ascidieria longifolia* (Hook. f.) Seidenfaden [Figs 98 & 99]

A highland epiphyte. Stems 3-25 cm long, tufted, slightly swollen at base, with numerous sheaths covering entire length of stem; two-leaved.

Leaves: Linear; acutely, unequally and convergently bilobed at apex; 12-25 cm long, 0.6-1.2 cm wide. No significant difference in size between leaves on single stem.

Inflorescences: One or two, axillary. Scape 4-6 cm long, rhachis 4-6 cm long, densely many-flowered. Scape, rhachis, pedicel + dorsal side of sepals densely pubescent. Scape with one large sheath and minute stem bracts (triangular-acute, to 1 mm long). Flowers in regular whorls with six flowers in each: not resupinate; white.

Floral bracts: Inconspicuous, triangular- acute, 0.5 mm long.

Pedicel + ovary: To 1 mm long.

Dorsal sepal: Ovate-subacute, 2.5 mm long, 1.5 mm wide.

Laterals: Obliquely triangular-subacute, 3.5 mm long, with mentum 1.5 mm long.

Petals: Obovate-subacute, slightly shorter than dorsal sepal.

Lip: Triangular-acute, widening from narrow base, 2 mm long, 1.5 mm wide (near apex).

Column and anther: White.

Distribution: Thailand, Malaysia, Sumatra.

Eria robusta (Bl.) Lindl. [Plate 3.3; Fig. 100]

Highland epiphyte, with two, much thickened, narrowly oblong-acute leaves, sometimes V-shaped in section or laterally compressed; up to 25 cm long, 2.5 cm wide; apical 1/4 recurved. Scape from base of stem, sturdy, 12-20 cm long, 6 mm wide, with fine, dense, brown hairs, and small stem bracts (triangular-acute, 2.5 mm long, 1.5 mm wide, scarious). Large, ovoid sheath at base of scape; 5.5 cm long, 2.2 cm wide.

Inflorescence: Densely many-flowered, 2.7 cm wide across the flowers. Rhachis, ovary and dorsal side of sepals with fine, dense, brown hairs.

Dorsal sepal: Oblong-acute, 4 mm long, 2 mm wide, pale yellow.

Laterals: Obliquely triangular-acute, 3.5 mm long, 5 mm wide at base, pale yellow, with joined bases forming mentum.

Fig. 101. *Eria pachystachya* Lindl. Alt. 1600 m. March.

Fig. 102. *Eria pachystachya*. 6.7×.

Fig. 103. *Eria densa* Ridl. Alt. 1600 m. September. 4.5×.

Petals: Obliquely ovate-obtuse, glabrous, with irregular margins, 3 mm long, 1 mm wide, pale greenish yellow.

Lip: Trilobed, 3.5 mm long, pale yellow. Side lobes erect and incurved, clasping column; 1 mm long, 2 mm wide, rounded. Mid-lobe triangular-acute, with decurved apex; 2 mm long, 1 mm wide.

Column: Flattened, widening from narrow base; apex pale green.

Anther: Purple.

Distribution: Thailand, Malaysia, Indonesia, New Caledonia, Fiji, Samoa.

Eria pachystachya Lindl. [Figs 101 & 102]

Large highland epiphyte. Stems 10-30 cm long, much thickened (1 cm or more), tapering towards base and apex.

Leaves: Mainly at apex of stem, longly lanceolate-acute, 5-20 cm long, up to 2 cm wide; with wide bases, tightly clasping stem.

Inflorescences: One or many, suberect, 6-12 cm long, up to 2 cm wide; densely many-flowered.

Bracts: Small, scarious, lanceolate-acute, 1.5 mm long.

Flowers: Small, not resupinate; very pale pink, with purple column and faint fragrance. Ovary, rhachis and dorsal side of tepals very finely and sparsely scurfy.

Dorsal sepal: Oblong-obtuse, 2.5-3 mm long, 1.5 mm wide.

Lateral sepals: Obliquely ovate-obtuse, 3.5 mm long, edges touching and in contact with lip; bases drawn out, forming a sac.

Lip: Spathulate, with wide (1.8 mm) blade and narrow (0.2 mm) claw. Whole lip 3-3.5 mm long.

Distribution: Thailand, Malaya, Sumatra, Java.

Eria densa Ridl. [Fig. 103]

Large highland epiphyte. Stems 10-30 cm long, to 3 cm thick; leafy in upper part; slightly elliptical in section.

Leaves: Oblong or narrowly elliptical, acute, to 30 cm long, 4 cm wide.

Inflorescences: Lateral, densely many-flowered, suberect. Rhachis slightly scurfy; pedicel and ovary with very sparse and short hairs. Flowers not resupinate.

Bracts: Reflexed, touching rhachis; oblanceolate-acute, 5 mm long, 1.5 mm wide; white, with brownish apex.

Dorsal sepal: Strongly decurved, covering column; ovate-acute, 4-6.5 mm long, 2.5 mm wide, white.

Lateral sepals: Obliquely triangular-acute, with wide bases forming mentum; 5 mm long, 5 mm wide at base; white, with faint purple flush on base.

Petals: Ovate-obtuse, 4.5 mm long, 2 mm wide, white, much incurved, covering column.

Lip: With wide, transversely elliptical blade from much narrower base. Small, acute, falcate lobes at base of lip. Whole lip 4.5 mm long; blade 3 mm wide, apex curved downward and inward. Lip purple, apex white.

Column: Short and thick; purple, with darker apex. Column-foot elongated.

Distribution: Thailand, Malaysia, Sumatra, Mentawi.

Eria elata Hook. f. [Figs 104 & 105]

Highland epiphyte. Stem short, thick, flattened, 4-10 cm long, up to 2.2 cm wide; with

Fig. 104. *Eria elata* Hook. f. Alt. 1600 m. 0.5×.

Fig. 105. *Eria elata*. 1.5×.

Plate 3
3.1. *Eria pulchella* Lindl. 4.5×.
3.2. *Eria leiophylla* Lindl. 3.5×.
3.3. *Eria robusta* (Bl.) Lindl. 5.3×.
3.4. *Bulbophyllum lepidum* (Bl.) J.J. Sm.
3.5. *Bulbophyllum uniflorum* (Bl.) Hassk. 1.9×.

3.1	3.2	
	3.3	
3.4	3.5	

3-5 large, acute, leaf-like, yellowish-green sheaths, laterally compressed in apical part, up to 11 cm long, 3.5 cm wide.

Leaves: 2-3, oblong to oblanceolate, acute; sheath 3-5 cm long, blade up to 19 cm long, 5 cm wide.

Inflorescences: 1-2, arising from leaf axils, up to 42 cm long, with flowers in apical part only.

Bracts: Large, conspicuous, ovate or broadly lanceolate, acute, up to 3 cm long, 1.5 cm wide, the lowest bract(s) without flower; green, flushed with purple, glabrous. Rhachis glabrous.

Pedicel + ovary: 5-11 mm long, with short brown hairs.

Tepals: Not much spreading.

Dorsal sepal: Oblong-acute, 12-21 mm long, 4-5 mm wide, pale (greenish) yellow; dorsal side with sparse, short brown hairs, sometimes barely visible.

Lateral sepals: Obliquely lanceolate-acute, pale greenish yellow, 10-13 mm long, 5-6 mm wide; bases joined to form short mentum, 4-6 mm long. Dorsal side with sparse, short brown hairs (not always apparent); mentum more densely hairy.

Petals: Obliquely lanceolate-acute, 9-14 mm long, 3-4 mm wide; white or very pale yellow.

Lip: Trilobed, attached to column-foot with very thin claw. Side lobes much larger than mid-lobe, rounded, erect, 3 mm long, 7 mm wide, with obtuse apex pointing forward; pale yellowish green. Mid-lobe triangular or rounded, retuse with very fine apiculus at nodal point; 2.0-2.5 mm long, 2.5-3.5 mm wide, (purplish) brown, with hairy callus in centre. Mesochilium with two cuneate keels, narrowing and diverging towards base. Base of side lobes with fine brown hairs on ventral side.

Column: Yellow, with winged sides; foot with reddish-brown, glandular hairs and tooth-like process.

Seed capsule: Prismatic, with marked wings; 3 cm long, 1 cm wide.

Eria lamonganensis Reichenb. f. [Figs 106 & 107]

A highland epiphyte. Base of stem creeping and rooting for about 2-4 cm, rest of stem semi-erect. New shoot branches off at point where preceding stem arises. Stems rather thick, 5-6 mm wide, 17-27 cm long; slightly widening (to 8 mm wide) in apical half; slightly tapering again towards apical tuft of 3-4 leaves. Internodes about 3 cm long. Stem generally covered with dry leaf-sheaths.

Leaves: Narrowly elliptic-acute, to 15 cm long, 3 cm wide, with marked mid-vein; rather dark green, upper surface glossy.

Inflorescence: From old leaf-sheaths in apical part of stem; many-flowered, 5 cm long.

Bracts: Conspicuous, elliptic-acute, 10 mm long, pale green, deflexed.

Pedicel + ovary: 11 mm long, with short, sparse, red hairs. Rhachis with short white hairs.

Dorsal sepal: Lanceolate-obtuse, 8 mm long, 3 mm wide.

Laterals: Obliquely lanceolate-acute, with wide bases, forming mentum; 8 mm long, 4 mm wide.

Petals: Lanceolate-acute, 7 mm long, 2 mm wide.

Tepals: All white, with longitudinal purple stripes. Mentum 4 mm long from onset of ovary; with sparse, short, red hairs.

Lip: 7 mm long, trilobed. Mid-lobe rounded, with irregular margins, yellow; divided into two, large, lateral lobes; V-shaped purple callus at base of mid-lobe, with short

Fig. 106. *Eria lamonganensis* Reichenb. f. Alt. 1060 m. October. 0.7×.

Fig. 107. *Eria lamonganensis*. 1.7×.

purple keel across callus. Side lobes erect, oblique, rounded at apex; 2 mm high, mauve with purple veins, margins nearly white. Blade of lip white, with pronounced, purple keel in centre, and two short keels on each side, appearing at base of lip and at base of mid-lobe.
Column: White, base pale yellow.
Anther: Yellow.
Distribution: Malaya, Java.

GENUS BULBOPHYLLUM Thouars, 1882

This strictly epiphytic, rarely lithophytic genus is probably the largest in the Orchid family, including more than 1000 species, distributed all over the tropics with a few representatives in sub-tropical Africa, Japan and New Zealand. South-East Asia is the area with most species: some 130 have been recorded for Malaya.

The plants have a creeping or hanging rhizome with a number of pseudobulbs. In a few cases, these are so small as to be easily overlooked, cf. section Aphanobulbum (page 91), but usually the pseudobulbs are quite conspicuous, doing justice to the name of the genus, from the Greek 'bolbos' = a bulb; 'phyllon' = a leaf. In Malaya, all *Bulbophylla* are one-leaved, the single leaf appearing at the top of the pseudobulb. The inflorescence arises from the base of the pseudobulb.

The flowers are usually quite small: some species of *Bulbophyllum* are among the smallest orchids known, as regards flower-size as well as the size of the plant (cf. *Bulbophyllum nematocaulon*, page 91). The sepals are generally larger, sometimes much larger, than the petals. The typical *Bulbophyllum*-lip is tongue-shaped and attached to the column-foot with a very thin claw. This arrangement makes the lip very easily movable, a way of attracting insects for pollination. The column is short, and usually provided with pointed arms, stelidia.

It is rare to find a *Bulbophyllum* in cultivation, as the flowers are small and not often scented. In fact, they are more often foul-smelling. A few large-flowered species, such as *Bulbophyllum macranthum* or *Bulbophyllum lobbii* (cf. page 87) are occasionally cultivated.

One large group within the genus *Bulbophyllum* is characterized by having the flowers in an apical, fan-shaped whorl, and by having the lateral sepals much longer than the dorsal sepal, with the margins more or less joined to form a synsepalum. This group, which is sometimes treated as a separate genus, is known as the section Cirrhopetalum (the reason for the generic name being that the dorsal sepal and the petals often have their edges fringed with conspicuous hairs). This section has a number of brightly-coloured, attractive species. The Malayan members of the section number about 28, including the first three species below.

Bulbophyllum gracillimum (Rolfe) Rolfe

A lowland epiphyte. Pseudobulbs spaced about 2 cm on rhizome; ovoid; about 2 cm long, 0.8 cm wide.
Leaf: Narrowly elliptical, emarginate; about 12 cm long; coriaceous. Scape very slender, 20-30 cm long; purple; many-flowered; the 8-14 flowers being in a whorl at the apex.
Dorsal sepal: Ovate and petals triangular, about 8 mm long, crimson; fringed with long hairs; apical bristle-like appendage about 5 mm long.

Fig. 108. *Bulbophyllum purpurascens* T. & B.
October. 2.3×.

Fig. 109. *Bulbophyllum purpurascens*. 5.5×.

Lateral sepals: Adnate at base, with long (5 cm), thread-like appendages.
Lip: Ligulate-rounded, thickened, decurved; about 3 mm long; violet-purple.
Distribution: Malaya, Indonesia, New Guinea.

Bulbophyllum purpurascens T. & B. [Figs 108 & 109]

A small epiphyte in lowland rain forest. Rhizome creeping, branching, with many roots. Pseudobulbs spaced about 1 cm; lenticular, with three sharp ridges; or prismatic-ovate, with four ridges; surface in line with leaf-blade often without a ridge. Size of pseudobulb 6 × 6 mm; base and part of one side adpressed to rhizome; green, flushed with purple.

Leaf: Elliptical (rarely oblanceolate or oblong), very slightly bilobed at rounded apex; very thick, lenticular in section, almost sessile; mid-vein depressed, conspicuous; 2-9 cm long, 1-3.5 cm wide. Scape 4.5-18 cm long, pale green with mauve sheaths at base, and one pale mauve sheath half-way between base and apex.

Inflorescence: 2-4-flowered, flowers arising at the same level from apex of stalk. Flowers rather unpleasantly scented (*Castanea sativa*-like scent).

Bracts: Lanceolate-acute, 3-3.5 mm long, pale mauve.

Pedicel + ovary: 3.5 mm long, pale green.

Dorsal sepal: Ovate-very acute; not spreading, but with a curved apex, forming hood over column together with petals; 5-6 mm long, 1.5-2 mm wide (when not flattened).

Lateral sepals: Linear-acute, base as wide as blade; 13-19 mm long, 2 mm wide, not much spreading.

Petals: Obliquely lanceolate-acute, with a few papillae on lower side, near base; 4.5-5 mm long, 1.3-2 mm wide; not spreading, but with recurved apices.

Tepals: Pale yellow; lateral sepals with raised, pale greenish mid-vein near apex.

Lip: Ligulate, rounded, with raised margins; much decurved, 2 mm long, 1 mm wide; pale yellowish green, lower side whitish.

Column: Short, white; rostellum linear, with triangular apex, pale green. Stelidia short, with two small teeth. Operculum very pale green.

Distribution: Thailand, Malaysia, Indonesia.

Bulbophyllum lepidum (Bl.) J.J. Sm. [Plate 3.4]

Epiphyte in lowland rain forest. Pseudobulbs ovoid-prismatic, close together, 2 cm long, 0.9 cm wide, covered with thin sheaths.

Leaf: Oblong-acute, apex decurved; base tapering into petiole; blade 12-16 cm long, 2.4-3 cm wide; petiole 2 cm long. Scape slender, 10-20 cm long, 1 mm wide, with one acute sheath, 6 mm long. Scape pale mauve, sheath pale yellow.

Inflorescence: In semi-circular umbel at apex of scape, 5-10-flowered. Flowers with faint and rather unpleasant scent.

Bracts: Longly acute, 4 mm long, very pale mauve.

Pedicel + ovary: 6 mm long, purple.

Dorsal sepal: Obovate-apiculate, convex, yellow, with purple base; 5-8 mm long, 3 mm wide; margins with long (1 mm) hairs.

Lateral sepals: Forming synsepalum, 25 mm long, 4-8 mm wide; apical part white, turning mauve and finally purple in basal part. Close to apex of synsepalum, the laterals are free from each other, and very slightly divergent.

Petals: Lanceolate, falcate, rounded; 4-6 mm long, very dark purple with hairy margins, not much spreading.

Fig. 110. *Bulbophyllum uniflorum* (Bl.) Hassk. Alt. 1500 m. December. 1.3×.

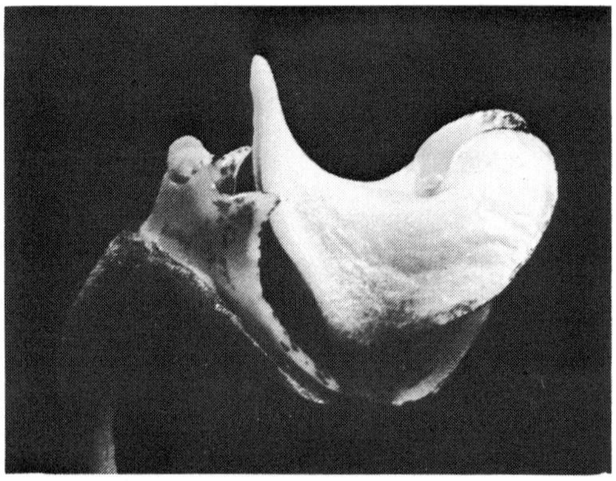

Fig. 111. *Bulbophyllum uniflorum*. Column with lip in normal position. 1.9×.

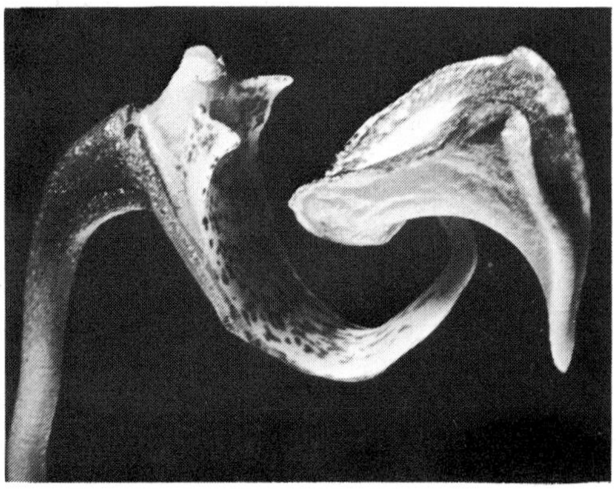

Fig. 112. *Bulbophyllum uniflorum*. Column with lip flapped down. 1.9×.

Fig. 113. *Bulbophyllum lobbii* Lindl. Alt. 1340 m. November.

Lip: Ligulate, acute, very mobile, decurved; apex yellow, base brown and V- shaped in section.
Column: Short; pale yellow flushed with purple in front. Operculum pale yellow.
Distribution: Malaysia, Indonesia.

Bulbophyllum uniflorum (Bl.) Hassk. [Plate 3.5; Figs 110, 111 & 112]
Highland epiphyte with long, sturdy rhizome, attached to substrate with numerous roots. Pseudobulbs widely spaced (5-12 cm), narrowly cylindrical, slightly flattened, often arching; to 10 cm long, 1.3 cm wide.
Leaf: Elliptic-acute; petiole 3-4 cm long; blade 15-20 cm long, 6-9 cm wide; mid-vein depressed, forming keel on lower side; apex decurved.
Inflorescence: With one or two flowers, only one open at a time; scape to 20 cm long, with a few acute sheaths, 1.9 cm long.
Floral bracts: Very broadly ovate-acute, 14 mm long; clasping pedicel; green, with purple veins.
Dorsal sepal: Lanceolate (widening from narrow base), longly acute; 40-50 mm long, 8 mm wide at base; 14-16 mm wide across blade; olive-green.
Lateral sepals: Obliquely lanceolate from wide base, longly acute, 40-50 mm long, 18 mm wide immediately above 8 mm wide basal attachment; olive green, ventral side with purple spots in basal half.
Petals: Lanceolate, longly acute, slightly oblique, 25-32 mm long, 11 mm wide near base; olive green, with pale mauve veins; apical part purple, but extreme tip (2 mm long) of slender apex light green.
Lip: Much thickened. Apical part much deflexed, base purple, apex and margins cream-coloured. Blade of lip with central cream-coloured spot; margins purple. Lower side of lip cream-coloured; margins of blade and basal part of apical lobe pale purple.
Claw: Yellow.
Column: Short, 6 mm long, pale green, with two short, erect stelidia, half as high as anther, with pointed, purple apices. Thickened, yellow wall round stigma. Column-wings drawn out to triangular, pointed processes on both sides of stigma. Processes 7 mm long (measured from ovary), white with purple spots. Column-foot very long, about 30 mm; twice bent through 90°; pale green, with purple spots. Operculum yellow.
Distribution: Malaya, Java, Sumatra.

Bulbophyllum lobbii Lindl. [Fig. 113]
A highland epiphyte. Rhizome thick (6 mm), branching, covered by bristly sheaths. Pseudobulbs ovoid, with three indistinct ridges; 3.5-5 cm long, 2.5-3 cm wide, yellowish green. Spacing on rhizome 2-8 cm.
Leaf: Oblong or oblanceolate, acute, apex decurved; tapering and conduplicate toward base; 17-25 cm long, 5-7 cm wide; mid-vein depressed.
Inflorescence: With single, large flower. The flower lasts up to a week, and has a faint, unpleasant, carrion-like scent.
Bracts: Ovate-acute, stiff, 15 mm long.
Pedicel + ovary: To 11 cm long, pale yellow or green, with mauve streaks.
Dorsal sepal: Lanceolate-acute, or apiculate, 32-50 mm long, 12-15 mm wide; pale yellow, densely spotted with dark purple; much spreading.
Lateral sepals: Lanceolate-acute, or apiculate, with wide bases joined to column-foot;

Fig. 114. *Bulbophyllum araniferum* Ridl. Alt. 1600 m. May. 0.7×.

Fig. 115. *Bulbophyllum araniferum*. 10.1×.

Fig. 116. *Bulbophyllum ovalifolium* (Bl.) Lindl. Alt. 1200 m. August. 5.7×.

Fig. 117. *Bulbophyllum ovalifolium*. Pseudobulbs and leaves. 4.3×.

Fig. 118. *Bulbophyllum ovalifolium* (f. flava). Alt. 1050 m. November. 2.5×.

Fig. 119. *Bulbophyllum ovalifolium* (f. flava). 6×.

falcate; 32 mm long, 16-18 mm wide at mauve base. Blade pale yellow, faintly spotted with mauve; much spreading; apex dark purplish brown on dorsal side.

Petals: Lanceolate-acute, or apiculate; much spreading or slightly recurved, slightly falcate, 25-29 mm long, 7-9 mm wide, pale yellow, with 5-7 purple veins.

Lip: Triangular-acute, cordate, much decurved, V-shaped in section, not much thickened; 13 mm long, 7-8 mm wide, pale yellow. Base with minute purple spots and short yellow keel above claw; apex with a few purple veins or spots. Margins of all tepals + lip often pure yellow, not spotted.

Column: Rather short, yellow. Column-foot very long, 15 mm, yellow, flushed with purple; apex curved sharply upwards.

Anther: Yellow.

Distribution: Assam, Burma, Thailand, Cambodia, Malaysia, Indonesia, Philippines.

Bulbophyllum araniferum Ridl. [Figs 114 & 115]

A highland epiphyte. Pseudobulbs narrowly ovoid, slightly flattened, horizontal with up-curved apex; 10-15 mm long, 7 mm wide; closely spaced, adpressed to rhizome.

Leaf: Lanceolate-acute, 3.5-6.5 cm long, up to 1 cm wide.

Inflorescence: 6-7 cm high, scape very slender, single-flowered.

Sepals: Longly lanceolate-acute, up to 30 mm long, base 3 mm wide; pale yellow with purple veins, apical part pale yellow.

Petals: Linear-obtuse, 3.5-5 mm long, 1 mm wide; pale purple with dark purple stripes.

Lip: Obovate, 10 mm long; deep purple. Base parallel with column-foot; margins erect. Blade deflexed, with sharply decurved margins. Apex tube-like.

Column: Pale purple, with long foot and two acuminate stelidia.

Distribution: Endemic?

Bulbophyllum ovalifolium (Bl.) Lindl. [Figs 116, 117, 118 & 119]

Very small epiphyte, found mainly in the highlands. Pseudobulbs in close contact, completely covering long, thin, branching rhizome; ovate to elliptical, slightly flattened and adpressed to rhizome; with slightly raised, rounded, bilobed apex; 3-6 mm high, 2.5-4 mm wide, pale green. A common variety has pale mauve tepals with dark mauve veins, a lip that is very dark purple, and a brownish mauve column.

Leaf: Oblong, oblanceolate or elliptic, rounded (occasionally unequally and connivently bilobed at apex), erect, 5-20 mm long, 3-7 mm wide; petiole 5-15 mm long.

Inflorescence: To 50 mm long, single-flowered; very slender, 0.2 mm wide. Scape 15-17 mm long, pale yellow.

Bracts: Small, 1 mm long; clasping stem, apex decurved.

Pedicel + ovary: 13 mm long. Flower not resupinate.

Dorsal sepal: Lanceolate-acute, not much spreading, 4.5-6 mm long, 2-2.5 mm wide; yellow, with pale orange veins.

Lateral sepals: Elliptic or obovate, (sub)acute, spreading; 7-10 mm long, 4-5 mm wide, orange-yellow with yellow veins.

Petals: Narrowly obovate-rounded, not much spreading, 1.2-1.5 mm long, 0.5 mm wide; yellow.

Lip: Ligulate, yellowish green. Apex thickened and much deflexed; margins papillose. Blade with two calli near base. Apical part of lip 1.5 mm long, 1 mm wide.

Column: Short, green; with two long, curving acuminate stelidia.

Distribution: Malaysia, Java.

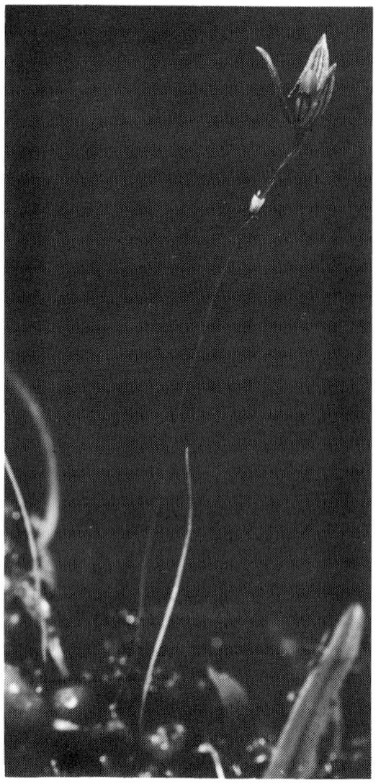

Fig. 120. *Bulbophyllum oeneum* Ridl. Alt. 1200 m. October. 1.8×.

Fig. 121. *Bulbophyllum oeneum.* 10.1×.

Fig. 122. *Bulbophyllum nematocaulon* Ridl. Alt. 1340 m. September. 4.7×.

Fig. 123. *Bulbophyllum mutabile* (Bl.) Lindl. Alt. 1340 m. November. 1.3×.

Fig. 124. *Bulbophyllum mutabile*. 6.7×.

Bulbophyllum oeneum Ridl. [Plate 4.1; Figs 120 & 121]

Very small highland epiphyte. Rhizome 0.7-1 mm wide, elongated, branching, with numerous roots. Pseudobulbs obliquely ovoid, base flattened and adpressed to rhizome; 6-8 mm long, 4 mm wide; spaced 6-20 mm.

Leaves: Oblong-apiculate, to 25 mm long, 3.5 mm wide; petiole less than 1 mm long; mid-vein depressed. Scape single-flowered, very slender; 50 mm long, 0.2 mm wide; dark purple, with one short, whitish, apiculate bract in apical part.

Pedicel + ovary: 6.5 mm long, pale green, flushed with purple.

Dorsal sepal: Oblong-very acute (or subacuminate), 5 mm long, 1.5 mm wide.

Lateral sepals: Obliquely lanceolate-acute, bases partly joined across column-foot; 6 mm long, 2.5 mm wide. All sepals pale yellow, with purple margins and three marked purple veins; apices pale yellow.

Petals: Obovate-obtuse, 2.5 mm long, 1 mm wide; pale yellow with purple margins and apex; marked, purple mid-vein.

Lip: Ligulate, much swollen, decurved; lateral margins erect. Whole lip 3 mm long, 1 mm wide (when not flattened), pale purple, papillose; apex rounded, pale yellow.

Column: Very short, pale green. Stelidia longer than column; incurved, white. Column-foot deep purple.

Anther: White.

Bulbophyllum nematocaulon Ridl. [Fig. 122]

Minute highland epiphyte, with long, slender, branching rhizome, forming dense mats. Pseudobulbs spaced about 10 mm, ovate, 2-4.5 mm high, 2-2.5 mm wide.

Leaf: Narrowly elliptic to oblong, acute or apiculate, to 20 mm long (often shorter, 5-7 mm long), 2-3 mm wide.

Inflorescences: Single-flowered, stalk + ovary 8-10 mm long.

Bract: 1 mm long, acute, scarious.

Sepals: All similar; lanceolate, tapering to a fine point; 6-7 mm long, 2 mm wide; pale green, spreading.

Petals: Obovate to narrowly elliptic; (sub)acute to apiculate; 2-3 mm long, 1-1.5 mm wide, pale green.

Lip: Oblong to triangular, obtuse; 4 mm long, 2 mm wide, green; blade slightly deflexed.

Column: Short, winged, pale green or white. Lateral wings rectangular, with fine points in corners.

Anther: Pale yellow.

Bulbophyllum mutabile (Bl.) Lindl. [Figs 123 & 124]

Rather small highland epiphyte with thin, branching rhizome, covered with sheaths throughout. Roots long, thin, numerous. No apparent pseudobulbs.

Leaf: Oblong to narrowly elliptic, obtuse, slightly thickened; to 45 mm long, 8 mm wide; petiole 3 mm long.

Inflorescences: 1-4-flowered.

Bracts: Acute, scarious, 8 mm long.

Flowers: Pale green or salmon-coloured.

Pedicel + ovary: 10 mm long.

Dorsal sepal: Lanceolate-acute, 5-6 mm long, 1.5-1.8 mm wide.

Lateral sepals: Obliquely lanceolate-acute, 4.5-5 mm long, with wide base.

Petals: Oblong or oblanceolate, obtuse, 3 mm long, 0.5 mm wide.

Fig. 125. *Bulbophyllum vaginulosum* Carr. Alt. 1650 m. November. 0.5×.

Fig. 126. *Bulbophyllum vaginulosum*. 5×.

Fig. 127. *Bulbophyllum brevipes* Ridl. Alt. 1600 m. July. 1.3×.

Fig. 128. *Bulbophyllum brevipes*. 7.9×.

Lip: Ligulate-rounded, much decurved, V-shaped in section; 4 mm long, 1.5 mm wide; darker green (or darker salmon-pink) than tepals.
Column: With two pointed, white stelidia.
Distribution: Malaysia, Indonesia.

Bulbophyllum vaginulosum Carr [Plate 4.2; Figs 125 & 126]
Highland and lowland epiphyte. Pseudobulbs small, not always apparent; closely spaced on rhizome; rounded, more or less vertically flattened; 3 mm long, 2.5-5 mm wide.
Leaf: Oblong or lanceolate, acute, tapering towards base; 7-25 cm long (including short petiole); 1.7-2.7 cm wide; thickened and stiff. Scape 3-10 cm long, covered with large, apiculate sheaths. Rhachis 7-18 cm long, rather sparsely many-flowered.
Bracts: Floral bracts lanceolate-acute, tightly clasping pedicel + ovary; 3-4 mm long; pale green, with light brown spots.
Flowers: White or pale yellow, with faint, rather unpleasant, *Castanea sativa*-like scent.
Pedicel + ovary: 2-3 mm long.
Dorsal sepal: Lanceolate-acute, widest in the middle, 6-10 mm long, 1.5-2 mm wide.
Laterals: Obliquely lanceolate, longly acute; with wide base, forming mentum; 7-10 mm long, 2 mm wide; mentum 1-2 mm long.
Petals: Narrowly (ob)lanceolate, acute or obtuse, 2.5-3 mm long, 0.3-0.7 mm wide.
Lip: 3.5 mm long, lanceolate-acute, blade deflexed, forming 90° angle with blade; blade 2-2.5 mm long, 0.7 mm wide, pale yellowish green; base with rounded, erect margins, 0.6-1 mm high. Blade with very finely hairy margins.
Column: With small, erect, straight, triangular to subulate stelidia; column-foot 1.5 mm long. Operculum yellow.

Bulbophyllum brevipes Ridl. [Plate 4.3; Figs 127 & 128]
Highland epiphyte with long, trailing, branching rhizome, often free from substrate. Pseudobulbs spaced 20-30 mm; ovate, with numerous ridges in old individuals; 15-30 mm long, 8-10 mm wide.
Leaf: Oblong to narrowly ovate, acute, 40-85 mm long, 15-22 mm wide; mid-vein depressed, forming keel on lower side. Petiole very short.
Inflorescences: One to many, 3-6-flowered; scape to 20 mm long.
Bracts: Lanceolate-acute, 3 mm long, with purple hairs (scurfy).
Tepals: Not much spreading.
Sepals: Lanceolate-acute, 10 mm long, 4 mm wide, pale yellow with darker tips.
Petals: Obovate-acute, sometimes with short apiculus; 4 mm long, 2.5 mm wide, very pale yellow.
Lip: Ligulate-obtuse, much thickened, decurved; 3 mm long when flattened, 2 mm wide, with two keels; pale yellow, with darker spot on hypochilium.
Column: White, with apiculate stelidia; column-foot long, slender, up-turned. Operculum white.
Distribution: Endemic?

Bulbophyllum sessile (König) J.J. Sm.
A similar, common species, with small pseudobulbs (3 mm high), and single-flowered, sheathed, very short-stemmed inflorescences, appearing in great profusion.

Fig. 129. *Bulbophyllum adenopetalum* Lindl. Alt. 1060 m. October. 0.9×.

Fig. 130. *Bulbophyllum adenopetalum*. 4×.

Fig. 131. *Bulbophyllum xanthum* Ridl. Alt. 800 m. December. 0.5×.

Fig. 133. *Bulbophyllum angustifolium* (Bl.) Lindl. Alt. 1400 m. 2.4×.

Fig. 132. *Bulbophyllum xanthum*. 10.8×.

Distribution: From Andaman Islands and Burma through Thailand and Malaysia to Indochina, Indonesia, New Guinea and islands in the Pacific.

Bulbophyllum adenopetalum Lindl. [Figs 129 & 130]
Highland and lowland epiphyte, with minute pseudobulbs, 3 mm high, 2.5 mm wide, closely spaced on rhizome, sometimes not apparent.
Leaf: Petiolate, very stiff and leathery, oblong-acute, 7-19 cm long, 0.8-2.5 cm wide. Petiole to 2 cm long. Scape to 10 cm long, with large, loose, pale green sheaths below rhachis. Rhachis 5-18 cm long, rather sparsely flowered.
Bracts: Lanceolate-acute, 4-6 mm long, pale green.
Flowers: Rather pale yellow, with *Castanea sativa*-like scent, not much opening.
Pedicel + ovary: 3 mm long.
Sepals: Lanceolate, very acute, 6-10 mm long, to 2 mm wide in the middle, 1 mm wide at base.
Laterals: Slightly oblique, with wide bases forming short mentum (2 mm long).
Petals: Oblanceolate-obtuse, 2.5-3 mm long, 0.5-0.7 mm wide.
Lip: Ligulate-rounded, trilobed, 3.5 mm long. Mid-lobe sharply deflexed, with hairy margins and some sparse white hairs on margin near base. Base loosely jointed to column-foot with narrow claw. Side-lobes semi-circular, erect, very close together, 1 mm long, 1.5 mm wide. No apparent keels, but basal part with slightly raised, yellow mid-vein.
Column: Short, white, with two acute arms, not much longer than anther. Operculum pale yellow.

Bulbophyllum xanthum Ridl. [Figs 131 & 132]
A lowland epiphyte with trailing, much branched, elongated rhizome, partly covered with sheaths; internodes about 2 cm long. Pseudobulbs narrowly conical, oblique, at 45° angle with rhizome, much wrinkled; to 2.8 cm long, 0.6 cm wide.
Leaf: Lanceolate-obtuse to subacute, to 12 cm long, 3.3 cm wide; mid-vein depressed; tapering toward base, forming short petiole.
Inflorescence: With several overlapping sheaths at the base; to 7 cm long; scape very short, 0.8 cm long; many-flowered. Flowers faintly scented; alternate on rhachis, facing the same way. Rhachis quadrangular, with sharp edges, and with sparse, very short hairs.
Bracts: Triangular-acute, scarious, 1.5 mm long.
Sepals: Oblong-acute, 6 mm long, 1.5 mm wide; apex recurved on dorsal sepal. Laterals slightly falcate.
Petals: Oblanceolate-rounded, 4 mm long. All tepals pale green.
Lip: Ligulate, decurved, thickened, pale yellow; with two lateral keels near margins. Apex rounded; blade 2.5 mm long.
Column: White, with two long, subulate-falcate arms, pointing forward.
Capsules: Rounded, diameter 12 mm.
Distribution: Endemic?

Bulbophyllum angustifolium (Bl.) Lindl. [Plate 4.4; Fig 133]
A highland epiphyte. Pseudobulbs well-spaced along rhizome; conical, 1-2 cm long, 5-8 mm wide.
Leaf: Lanceolate-acute, 2.5-9 cm long, 0.9-1.5 cm wide.
Inflorescences: 2-5-flowered; with short stalk, 5 mm long.

Fig. 134. *Bulbophyllum parvilabium* Schltr. Alt. 1450 m. September. 0.8×.

Fig. 135. *Bulbophyllum parvilabium*. 4.8×.

Fig. 136. *Bulbophyllum foetidolens* Carr. Alt. 1400 m. December. 0.7×.

Fig. 137. *Bulbophyllum foetidolens*. 6.4×.

Plate 4
4.1. *Bulbophyllum oeneum* Ridl. 7.6×.
4.2. *Bulbophyllum vaginulosum* Carr. 4.6×.
4.3. *Bulbophyllum brevipes* Ridl. 4.3×.
4.4. *Bulbophyllum angustifolium* (Bl.) Lindl. 6.8×.
4.5. *Bulbophyllum parvilabium* Schltr. 7.6×.

4.1	4.2	4.3
4.4		4.5

PLATE 4

Bracts: Lanceolate-acute, 2 mm long.
Flowers: Bright orange; base of tepals pale yellow.
Sepals: Lanceolate-longly acute, 6-10 mm long, laterals very slightly longer.
Petals: Obovate, obtuse to subacute, 2.5-5 mm long.
Lip: 1.5-2 mm long; ligulate, with sharply reflexed apex.
Distribution: Malaya, Java.

Bulbophyllum parvilabium Schltr [Plate 4.5; Figs 134 & 135]
Highland epiphyte with elongated rhizome, often pendulous and not in contact with substrate. Pseudobulbs very closely spaced; parallel with, and close to rhizome; narrowly ovate; to 2 cm long, 0.6 cm wide.
Leaf: Narrowly elliptic-obtuse (sometimes apiculate, or equally & obtusely bilobed at apex), to 7.5 cm long, 1.8-2 cm wide.
Inflorescences: With short stalks, 5-10 mm long; about 3-flowered.
Bracts: Thin, lanceolate-acuminate, 3.5 mm long.
Sepals: Lanceolate-acute from wide base; 7 mm long, 3 mm wide at base; bright orange at apex, very pale yellow at base.
Petals: Ovate-apiculate, 3 mm long, 1.5 mm wide; apex pale orange, base white.
Lip: Ligulate, with decurved, triangular apex, 1 × 1 mm (not flattened). Margins of hypochilium recurved; base much thickened. Whole lip orange, with darker apex.
Column: Short, blunt; with two subulate, erect stelidia, nearly 1 mm higher than anther. Column-foot long, curved strongly backwards. Column and operculum white.
Distribution: Malaya, Sumatra.

Bulbophyllum foetidolens Carr [Figs 136 & 137]
A highland epiphyte. Rhizome 2-4 mm wide, elongated, developing many roots at 10 mm-intervals. Pseudobulbs widely spaced (4-12 cm); pyriform, with wide base adpressed to rhizome; 2.5-5 cm high, 0.6-2.3 cm wide; green.
Leaf: Oblong-obtuse (or equally and obtusely bilobed at apex); 15-20 cm long, to 3.8 cm wide; slightly stiff and thickened; tapering towards base, forming short (5 mm) petiole; mid-vein depressed, forming keel on lower side. Scape slender, elongated, dark purple, 15-35 cm long, 1 mm wide; with 6 clasping, 1.6 cm long sheaths. Rhachis 5-10 cm long, finely warty, more or less pendulous, purple; very densely many-flowered, all flowers open at the same time. Flowers with faint, unpleasant scent.
Bracts: Lanceolate-acute, 2-4 mm long, pale yellow with two purple veins.
Pedicel + ovary: 3.5-7 mm long, with irregular, finely warty surface; purple.
Dorsal sepal: Lanceolate-acute, widest in the middle, 4-7 mm long, 2-3 mm wide; base dark purple, blade pale yellow, with three, dark purple veins; dorsal side finely purple-hairy.
Lateral sepals: 5-7 mm long, 2-3 mm wide, very oblique, pale yellow with 4-5 purple veins; dorsal side finely purple-hairy. Apices connivent.
Petals: Narrowly oblong-rounded, with small apiculus; 2-5 mm long, 0.3-1 mm wide; pale yellow with purple margins and mid-vein.
Lip: Much swollen, 2-3 mm long. Blade ligulate, decurved, pale yellowish-green; with large cushion, projected backward into two narrow keels, ending in purple points at base of lip. Margins of lip forming another two purple points, near ends of keels. Whole base of lip + claw purple.

Fig. 139. *Bulbophyllum tahanense*. 12×.

Fig. 138. *Bulbophyllum tahanense* Carr. Alt. 1650 m. September. 0.5×.

Fig. 141. *Agrostophyllum bicuspidatum*. Stem with leaves. 3.6×.

Fig. 140. *Agrostophyllum bicuspidatum* J.J. Sm. Alt. 1400 m. July. 13.5×.

Column: Purple, with two small, dark purple teeth facing lip, and two long, subulate, pale yellow stelidia, erect above anther; one larger, purple tooth at base of column; column-foot purple. Operculum pale yellow and purple.

Distribution: Probably endemic.

Bulbophyllum tahanense Carr [Figs 138 & 139]

A small highland epiphyte. Rhizome rather thick, 3-3.5 mm wide, completely covered with sheaths. Internodes 5-15 mm long. Pseudobulbs obliquely conical, with wide base adpressed to rhizome; apex upturned; 15-20 mm long, 12 mm wide; yellowish green.

Leaf: Oblong-rounded, 23-80 mm long, 9-17 mm wide; stiff, erect; sometimes equally bilobed at apex. Petiole short, 2-4 mm long. Scape much longer than pseudobulb + leaf; to 15 cm long, slender (0.5 mm wide), dark brownish purple, with a few acute sheaths. Rhachis 3-7 cm long, about 10-flowered.

Bracts: Acute, 1.2 mm long.

Pedicel + ovary: 7-10 mm long.

Dorsal sepal: Lanceolate-acute, 7.5-10 mm long, 3 mm wide; pale yellow; not much spreading.

Lateral sepals: Obliquely lanceolate-acute with wide base; 7-9 mm long, 3 mm wide; pale yellow, apex tinged with mauve; apical part spreading.

Petals: Ovate-acute, 5 mm long, 2-3 mm wide; very pale yellow, not spreading.

Lip: Loosely attached to column-foot, triangular-subacute, much decurved; base of lip with two high keels at 90° angle to each other, giving lip a V-shaped appearance. Blade of lip with single, slender, papillose callus. Whole lip purple.

Column: Short, white, with very short stelidia. Operculum cream-coloured, with small triangular point, projected forwards.

Distribution: Probably endemic.

GENUS AGROSTOPHYLLUM Blume, 1825

The genus comprises plants with two distinct types of growth: the first type having slender stems with many internodes and numerous small leaves at right angles to the stem; the inflorescences have one or a few flowers open at a time. The other type has much longer (15-100 cm) leaves at an acute angle to the stem, and many-flowered, head-like inflorescences. Both types have 8 pollinia, attached to a single disc. Of the 60 species of this South Asian genus, seven are found in Malaya. The generic name means 'grass-leaved', from Gk 'agrostis' = a kind of grass, and 'phyllon' = a leaf.

Agrostophyllum bicuspidatum J.J. Sm. [Figs 140 & 141]

Common lowland and highland epiphyte. Stems numerous, stiff, tufted, often leafy throughout; up to 50 cm long, slender. Internodes about 5 mm long.

Leaves: Numerous, at right angles to the stem; with very short (0.3 mm) petiole. Blade lanceolate, hastate; equally (or subequally) and obtusely bilobed at apex; up to 1.2 cm long, 4 mm wide. Leaf-blade with circular depressions, ∅1 mm, three on each side of mid-vein, seen in raised relief on lower side, giving the plant a fern-like appearance. Leaf-sheaths with cuspidate appendage, up to 3 mm long, one on each side of the stem.

Inflorescence: Terminal (rarely lateral), arising from tuft of scarious bracts; many-flowered, only a few flowers appearing at a time. Flowers yellow.

Fig. 142. *Agrostophyllum glumaceum* Hook. f. August. 0.7×.

Fig. 143. *Agrostophyllum glumaceum*. 7.1×.

Fig. 145. *Ceratostylis ampullacea*. 9×.

Fig. 144. *Ceratostylis ampullacea* Kränzl. Alt. 1650 m. May. 2.6×.

Dorsal sepal: Oblong-acute, 4 mm long, 2 mm wide.
Lateral sepals: 4 mm long, 4 mm wide at base; obliquely triangular, with wide base forming mentum.
Petals: Narrowly lanceolate-acute, 3 mm long, 0.6 mm wide.
Lip: Trilobed with saccate base; blade 4 mm long. Mid-lobe semi-circular, obtuse, pale yellow; 1.5 mm long, 2 mm wide. Side lobes small, erect, white. Sac 1.5 mm wide, hidden in mentum; white.
Column: 2 mm long, pale green, with purple apex.
Distribution: Burma, Thailand, Malaya, Indonesia, Solomon Islands.

Agrostophyllum glumaceum Hook. f. [Figs 142 & 143]
Large lowland epiphyte. Stems numerous, tufted, thickened, much flattened; to 20 cm long, slightly tapering from wider (2.5 cm) base; with one short, apical node, 1-6 cm long.
Leaves: Four, 2 basal and 2 apical ones; linear, 30-50 cm long, to 2 cm wide; very acutely and very unequally bilobed at apex; apical leaves longer than basal ones. All leaves with long, flattened sheaths, to 8 cm long on basal leaves, very short on topmost of apical leaves.
Inflorescence: Terminal, branched, densely many-flowered, flowers forming head. Rhachis zig-zag.
Bracts: Acute, up to 5 mm long.
Pedicel + ovary: 6 mm long, sparsely scurfy.
Sepals: White; dorsal side slightly scurfy.
Dorsal sepal: Ovate-acute, not much spreading, 5 mm long, 2 mm wide.
Lateral sepals: Ovate-acuminate, 6 mm long, 2.2 mm wide, forming short mentum; not much spreading.
Petals: Linear, acute, arching, 4 mm long, 0.5 mm wide, white.
Lip: Trilobed with saccate base. Side lobes small, rounded, erect, enclosing base of column. Mid-lobe rhomboid, 4 mm long, 3.5 mm wide, with irregularly dentate margin; small teeth near apex, larger teeth at base. Blade of lip very pale yellow, with very pale mauve spots.
Distribution: Malaysia, Indonesia.

GENUS CERATOSTYLIS Blume, 1825

A genus of about 60 species in South Asia, with 12 species found in Malaya. Many species are very strange-looking and when not flowering, they seem more like rushes than orchids. These rush-like orchids have a single, more or less terete leaf, which is very similar to the stem. The leaf continues the line of the stem above the terminal inflorescence. The stems are densely tufted. In other species, the leaf is not terete, and the stem is typically covered with dry sheaths. Etymology: Gk 'keras' = horn; 'stylos' = column.

Ceratostylis ampullacea Kränzl. [Plate 5.1; Figs 144 & 145]
Highland epiphyte with tufted, green, rush-like stems, 15-30 cm long, 2-3 mm wide.
Leaf: Single at end of stem; terete-acute, with faintly marked longitudinal keel, facing the inflorescence; 2.5-5 cm (most commonly 3-3.5 cm) long.
Inflorescence: Terminal, arising from the same joint as leaf, among small, chaffy bracts; 2-3 flowers open at the same time.

Fig. 146. *Ceratostylis gracilis* Bl. Alt. 1340 m. September.

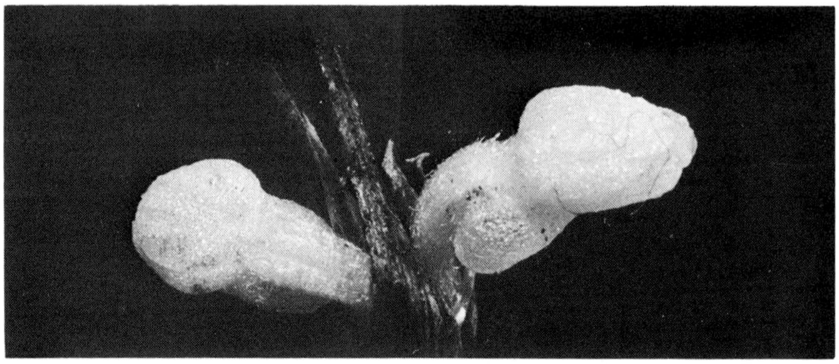

Fig. 147. *Ceratostylis gracilis*. 9×.

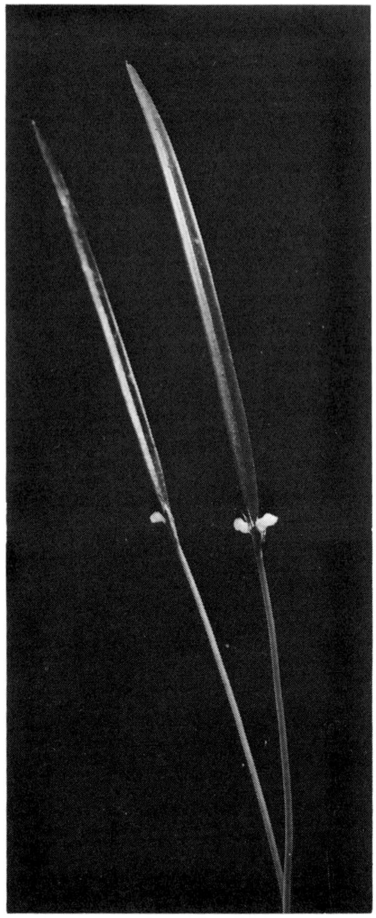

Fig. 148. *Ceratostylis gracilis*. 1×.

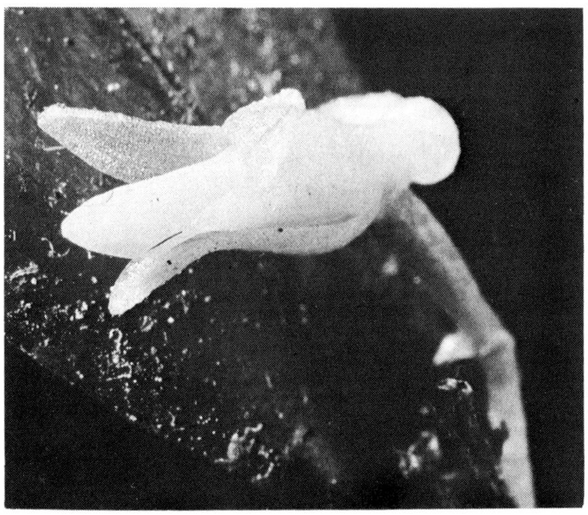

Fig. 150. *Ceratostylis lancifolia*. 7.8×.

Fig. 149. *Ceratostylis lancifolia* Hook. f. Alt. 1600 m. June. 1.1×.

Pedicel + ovary: 5 mm long, purple with white hairs.
Dorsal sepal: Lanceolate-subacute, apex recurved; 2.5-3 mm long, 1.2 mm wide; pink to pale purple.
Lateral sepals: Distal part similar to dorsal sepal; basal part forming pouch-like mentum, much swollen and slightly flattened in apical part. Length of mentum 2-3.5 mm, width at apex 2 mm; pink to purple, with sparse white hairs.
Petals: Narrowly lanceolate-acuminate, 2.5 mm long, pale pink. All tepals recurved at apex.
Lip: Ligulate-obtuse, indistinctly trilobed. Side lobes close to column, white. Mid-lobe much swollen, 1.5 mm long, yellow. Operculum white.
Distribution: Thailand, Malaya, Sumatra.

Ceratostylis gracilis Bl. [Figs 146, 147 & 148]
A highland epiphyte. Stems numerous, densely tufted, slender, erect; with small sheath (2 cm long) at base; to 8 cm long, 1.1 mm wide, single-leaved.
Leaf: Lanceolate-very acute, thickened, with depressed mid-vein; to 8.5 cm long, 3 mm wide.
Inflorescence: From tuft of dry bracts, 1-2-flowered.
Bracts: Acuminate, 8 mm long, scarious.
Flowers: White, not opening.
Pedicel + ovary: 4 mm long, sparsely and finely hairy.
Dorsal sepal: Ovate-rounded, 2 mm long, 1.5 mm wide.
Lateral sepals: 4 mm long, bases forming mentum.
Petals: Linear, 2 mm long.
Lip: 3 mm long, widening from narrow base; apical part triangular, swollen. Apex slightly diffused with mauve.
Column: Short, white; with two porrect, subspathulate arms, much longer than anther. Operculum white.
Distribution: Malaya, Sumatra, Java.

Ceratostylis lancifolia Hook. f. [Figs 149 & 150]
Small highland epiphyte. Stems slender, hardly swollen at base, 1-leaved; 5-10 cm high, often entirely covered with scarious sheaths.
Leaf: Narrowly elliptical to oblong, acute, stiff and leathery, 4-7.5 cm long, 0.9-1.3 cm wide.
Inflorescence: From tuft of withered bracts in leaf axil; only one or two flowers open at the same time.
Pedicel + ovary: 1.5 cm long, with very short, white hairs.
Dorsal sepal: Linear-acute, 7 mm long, 1.5 mm wide, apex slightly recurved.
Lateral sepals: Lanceolate-acute, joined at broad bases, 8 mm long, 2 mm wide, forming pouch-like, slightly bilobed mentum, 2 mm long, 2 mm wide.
Petals: Lanceolate-acute, 5 mm long, 1 mm wide.
Lip: Ligulate-obtuse, trilobed, 7 mm long. Mid-lobe very swollen. Side lobes indistinct, erect. Lip encloses column completely. All tepals pale green; mid-lobe of lip pale yellow.
Column: With two ovoid, porrect stelidia.

Ceratostylis eriaeoides Hook. f. [Figs 151 & 152]
Small, attractive highland epiphyte. Stems close together, short (to 1 cm), single-

Fig. 151. *Ceratostylis eriaeoides* Hook. f. Alt. 1650 m. August. 1×.

Fig. 152. *Ceratostylis eriaeoides*. 3.6×.

Fig. 153. *Ceratostylis clathrata* Hook. f. Alt. 1250 m. September. 3.5×.

Fig. 154. *Ceratostylis* sp. (Segerbäck 2145). Alt. 1100 m. December. 1.1×.

leaved, complctely covered with scarious sheaths; not much swollen at base.

Leaf: Obovate or oblanceolate to elliptical; very slightly obtusely & unequally bilobed at apex; 3-5 cm long, 0.5-1.4 cm wide, very thick and stiff; semi-lenticular in section (one side is flattened); petiole 1 cm long. Scape to 2.5 cm long, very pale green, puberulent.

Bracts: Acute, 3 mm long.

Flower: Single, white.

Pedicel + ovary: To 6 mm long, puberulent with hairs adpressed to stem.

Dorsal sepal: Narrowly ovoid-(sub)acute, 10-14 mm long, 4-5 mm wide, slightly recurved.

Lateral sepals: Obliquely triangular-acute, 12-14 mm long, 5 mm wide at base; joined at base, forming short mentum, 1.5 mm long from onset of ovary; mentum very pale green.

Petals: Elliptical-acute, falcate, 12 mm long, 3.5-4 mm wide.

Lip: Very indistinctly trilobed. Side lobes rounded, erect, pale green, very small. Mid-lobe triangular, 3 mm long, 2.5 mm wide, swollen, green. Base of lip pale green, saccate, with two dark brown keels from base of mid-lobe to opening of spur. Sac 2 mm long, pale green, hidden by mentum.

Column: White; with long, pale green foot, and two large, obovate processes, 1.5 mm long. Operculum white.

Ceratostylis clathrata Hook. f. [Fig. 153]

Small highland epiphyte with trailing rhizome, entirely covered with reticulate sheaths. Stems to 4 cm long, branching, covered with reticulate sheaths.

Leaves: Often in fan-like arrangement; much swollen, terete or nearly terete; to 3 cm long (often shorter), 3 mm wide, acute, sometimes very slightly and unequally bilobed at apex; base covered with reticulate sheath.

Inflorescence: From leaf-axil, single-flowered. Flowers not much opening.

Pedicel + ovary: To 8 mm long.

Dorsal sepal: Lanceolate-acute, 3.5-4 mm long, 1.5-2 mm wide; pale yellow with mauve veins.

Lateral sepals: Triangular-acute, 3 mm long, 3 mm wide at base; mentum 2 mm long from onset of ovary; pale yellow with mauve veins.

Petals: Narrowly lanceolate-acute, 2 mm long, pale mauve.

Lip: With triangular, deflexed apex, 1 mm long.

Column: Mauve.

Anther: White.

Distribution: Probably endemic.

Ceratostylis sp. (Segerbäck 2145) [Fig. 154]

A species near *Ceratostylis* No. 8 in Holttum (1972). It is a small highland epiphyte with very short, densely tufted stems and 1-3 leaves.

Leaves: Stiff, slightly thickened, oblong; mid-vein depressed; narrowing at base, very unequally bilobed at apex; to 6 cm long, 4.5 cm wide. Leaf-bases and stem covered with dry sheaths.

Inflorescence: Single-flowered, flower white.

Bracts: 6 mm long, scarious.

Pedicel + ovary: To 2.3 cm long, with dense, short, white hairs.

Dorsal sepal: Lanceolate-acute, 8 mm long, 1.2 mm wide, hairy on dorsal side.

Fig. 155. *Podochilus microphyllus* Lindl. Alt. 1450 m. March. 7.1×.

Fig. 156. *Podochilus muricatus* (T. & B.) Schltr. Alt. 800 m. May. 0.3×.

Fig. 157. *Podochilus muricatus*. 3.6×.

Lateral sepals: Lanceolate-acute, 10 mm long; with wide base forming mentum, 2 mm long. Mentum rounded, pointing backwards & parallel with ovary. Lateral sepals hairy on dorsal side.

Petals: Lanceolate-longly acute, 8 mm long, 0.6 mm wide, glabrous.

Lip: 8 mm long, attached to column-foot; widening from narrow, tubular base; blade lanceolate-acute, at 90° with base. Basal part of blade with erect, rounded margins.

Column: White; with two oblique, rounded, erect processes, longer than anther.

GENUS PODOCHILUS Blume, 1825

A South Asian epiphytic genus with four species in Malaya. Two of these have very small leaves, giving the short plants a moss-like appearance. The flowers are small and white, often with purple spots. The lateral sepals form a blunt mentum, and the lip is provided with small appendages at the base. Gk 'pous' = foot; 'kheilos' = lip.

Podochilus microphyllus Lindl. [Fig. 155]

A small lowland and highland epiphyte. Stems numerous, generally leafy throughout; often branched; 5-20 cm long.

Leaves: Numerous, ovate-apiculate, 5-8 mm long, 2-3 mm wide; twisted at base.

Inflorescences: Usually at apex of leafy stalks, 5-10-flowered. Rhachis pale purple. Flowers not resupinate.

Bracts: Lanceolate to triangular, acute, 1 mm long.

Dorsal sepal: Ovate-obtuse, 2 mm long, 1.5 mm wide; white, with purple streak in the middle.

Lateral sepals: Triangular-obtuse, with base projected backwards to form two pouch-like structures, 3.5 mm long; laterals white, with longitudinal purple streak in the middle.

Petals: Narrowly ovate-obtuse, slightly longer than dorsal sepal; white, with longitudinal purple streak in the middle.

Lip: Ligulate-subacute, slightly decurved, 2 mm long; base with short, wide appendage; apex purple.

Column: Short, pale green; apex of rostellum often with purple spot.

Distribution: Burma, Cambodia, Thailand, Vietnam, Malaysia, Indonesia.

Podochilus muricatus (T. & B.) Schltr [Figs 156 & 157]

Small mountain epiphyte, with a few to numerous, tufted stems, up to 15 cm long, leafy throughout.

Leaves: Obovate, to 12 mm long, 7 mm wide, unequally bilobed at apex; lobes rounded, with fine apiculus between lobes (prolongation of mid-vein); leaves tightly spaced, 2-ranked; blades twisted at base and forming one plane. Leaves sheathed, rather stiff; green, sometimes with purplish flush.

Inflorescences: Terminal, scape very short. Rhachis 7 mm long, covered with bracts, the old ones chaffy.

Bracts: Broadly obovate-apiculate, 3 mm long, 2.5 mm wide, pale green, hyaline.

Flowers: White, only a few open at the same time.

Sepals: Lanceolate-acute, 7 mm long, very longly white-papillose on dorsal side; apices greenish; not much opening.

Petals: Lanceolate-acute, glabrous, shorter than sepals; white, apex pale green, mid-vein purple (esp. on ventral side).

Fig. 158. *Appendicula ovalis* (Schltr) J.J. Sm. Alt. 1050 m. November. 0.7×.

Fig. 159. *Appendicula ovalis*. 9.4×.

Fig. 160. *Appendicula undulata* Bl. January.

Fig. 161. *Appendicula undulata*. 11.6×.

Lip: 5 mm long, 2.5 mm wide (blade); widening from narrow base; apex with small, tooth-like callus; blade with two large, diffuse, purple spots, and two smaller, distinct spots closer to base; small, purple, tooth-like appendage near base.
Column: Purple in front; with two, pale purple, tooth-like, erect arms (slightly diverging). Rostellum elongated, 3 mm long, parallel with petals.
Seed capsules: Densely papillose, less than 10 mm long.
Distribution: Malaya, Sumatra, Java.

GENUS APPENDICULA Blume, 1825

The genus is named after an appendage on the upper surface of the lip, near the base (in Latin, 'appendicula' means 'a small appendage'). The species often have long, slender, pendulous stems, with numerous, 2-ranked leaves; these are twisted at the base, so that the leaf-blades are all in one plane. The flowers are small, with a distinct mentum.

Appendicula ovalis (Schltr) J.J. Sm. [Figs 158 & 159]
A highland epiphyte. Stems pendulous, branching, 50 cm long, or more; often leafy throughout. Internodes 4 mm long.
Leaves: Numerous, elliptic-rounded; largest leaves 11-18 mm long, 6-9 mm wide, leaf-size much varying on same individual. Apex of leaf retuse, with very fine point (prolongation of mid-vein).
Inflorescences: Lateral and terminal, to 10 mm long, covered with sheaths; single-flowered.
Bracts: Sheathing rhachis; acute, scarious, to 4 mm long; apex deflexed.
Pedicel + ovary: 4 mm long.
Dorsal sepal: Lanceolate-acute, 3.5 mm long, 2 mm wide; white, flushed with purple (esp. mid-vein).
Lateral sepals: Obliquely triangular-acute, 5 mm long, 2.5 mm wide (across blade), with very wide bases, joined in front of lip-base, forming mentum; mid-vein raised on dorsal side. Blade very pale purple, with darker purple marking on dorsal side of blade. Mentum white, saccate, 3 mm long, 2.5 mm wide; at right angles to ovary.
Lip: With blade 4 mm long, white; apex 2.5 mm long, twisted to the right, purple, sharply deflexed. Base and blade with 2.5 mm long, concave appendage with rounded apex, starting half-way up the blade, ending at base of lip.
Column: White, with purple apex.
Distribution: Malaya, Thailand, Sumatra, Java.

Appendicula undulata Bl. [Figs 160 & 161]
A lowland epiphyte. Stems elongated, erect when young, later pendulous; branching only at base; 1 m long, or more, leafy throughout.
Leaves: Narrowly elliptical in one plane, 2-ranked; margins slightly overlapping; with very long sheaths, covering stem; apex deflexed, forming apiculus; blade 1.9-3.5 cm long, 1.1-1.8 cm wide.
Inflorescences: Axillary, slender, pendulous, from lower side of stem; scape to 2 cm long, covered with very acute, brownish mauve sheaths, 4 mm long, overlapping.
Bracts: Lanceolate-acute, 4 mm long, 2 mm wide, pale mauve, spotted with purple.
Flowers: 1-2, small, 7 mm long from end of mentum to apex of dorsal sepal.

Fig. 162. *Appendicula reflexa* Bl. January. 1.3×.

Fig. 163. *Appendicula reflexa*. 7.5×.

Pedicel + ovary: 2.5 mm long.

Dorsal sepal: Broadly ovate-acute, 3.5 mm long, 3 mm wide, pale mauve with darker mid-vein; apex recurved.

Lateral sepals: Ovate-acute (blade), with very wide bases, joined at the edges, forming mentum. Blade 3.5 mm long, mentum 3 mm long. Blade pale mauve with darker apex and mid-vein. Mentum white.

Petals: Oblanceolate-obtuse, 2.1 mm long, 1.9 mm wide. Base white, apical half purple, with white mid-vein.

Lip: 4 mm long, 2 mm wide; gradually widened from narrow base; apical part purple, deflexed; base of lip thickened (appendage).

Column: Pale green.

Distribution: Malaysia, Thailand, Sumatra, Java.

Appendicula reflexa Bl. [Figs 162 & 163]

Lowland and mountain epiphyte. Stems elongated, branching only at base, pendulous; round in section, leafy throughout; to 1 m long.

Leaves: Oblong to elliptical (often slightly wider near base); rounded, with deep notch at apex and sharp tooth between the unequal lobes; mid-vein depressed, forming keel on lower side; leaf-bases sheathing stem; leaves 2-ranked, more or less in one plane; blade to 4.5 cm long, 1.5-2 cm wide, often flushed with purple. Internodes 8-10 mm long; leaves close, but generally not overlapping at edges.

Inflorescences: Usually lateral; many-flowered. Scape very short, 5 mm long; rhachis 10-20 mm long, often purple; inflorescences appearing on lower side of stem, with flowers in dense heads (about 10 flowers). Most of the flowers are open at the same time.

Bracts: Lanceolate-acute, rather persistent, green or purple, deflexed, 2-3 mm long, 1.5 mm wide.

Pedicel + ovary: 2-3 mm long.

Dorsal sepal: Triangular to ovate, obtuse to rounded; 2.5-3 mm long, 1.5 mm wide; not much spreading; green, often flushed with purple.

Lateral sepals: Triangular to ovate, obtuse, with wide base, forming mentum; blade 2.5 mm long; mentum sac-like, 1.5-2 mm long. Laterals pale green, mentum almost white.

Petals: Lanceolate, obtuse or rounded, 2 mm long, 0.7-1 mm wide, not much spreading; very pale green.

Lip: Uppermost, 3 mm long, 2 mm wide, blade obcordate-obtuse, decurved, white; pale green mid-vein forms a slightly thickened point at apex; appendix at base of lip.

Column: Green, winged at apex.

Anther: Yellowish.

Distribution: Thailand, Vietnam, Malaysia, Indonesia, New Guinea, New Caledonia, Solomon Islands, New Hebrides, New Ireland, Fiji.

Appendicula sp. (Segerbäck 2153)

Lowland epiphyte. Stems elongated, flattened, branching; to 30 cm long; winged (due to leaf-bases); more or less pendulous; leafy throughout.

Leaves: Numerous, 2-ranked, in one plane; at about 45° with stem; obovate, more or less equally bilobed at apex; lobes rounded; rather pale green. Edges of leaves overlapping; sheaths imbricate; internodes 3 mm long. Leaves with conspicuous

Fig. 164. *Dipodium pictum* (Lindl.) Reichenb. f. Alt. 1100 m. December. 0.5×.

Fig. 165. *Dipodium pictum* (Lindl.) Reichenb. f. 2.2×.

venation. Extension of mid-vein forms sharp tooth between lobes at apex. Length of leaves 10 mm, width 5 mm.

Inflorescence: Terminal, from tuft of white, leaf-like bracts.

Bracts: Broadly lanceolate, acute or apiculate, white, with green apiculus; laterally compressed, curving downwards, 8 mm long, 6 mm wide, much longer than tepals; with much narrower base (1 mm).

Flowers: White, 6 mm long.

Dorsal sepal: Ovate-apiculate, much convex, 2.5 mm long (excl. apiculus), 1.5 mm wide (not flattened). Not spreading (in line with ovary).

Lateral sepals: Ovate, 5 mm long, 2 mm wide (not flattened); with long (0.7 mm) apiculus; bases joined in lower half, forming large pouch, 2 mm long, 2.5 mm wide; distal half with raised mid-vein forming keel on dorsal side and ending with apiculus; not spreading.

Petals: Ovate-obtuse, 2 mm long, 1.5 mm wide, not spreading.

Lip: 5 mm long, trilobed. Mid-lobe retuse and fimbriate at apex; 1.5 mm long, 2.5 mm wide; side lobes small, erect, rounded. Apex of lip twisted towards the right-hand side.

GENUS DIPODIUM R. Brown, 1810

A small genus (about 20 species) with three species recorded from Malaya. Two of these are climbing epiphytes, one is a terrestrial. The stems are rather long, with long, arching, narrow leaves. The inflorescences are many-flowered, with rather large flowers. Sepals and petals are pale yellow, with large reddish spots, especially on the dorsal side. The lip (or part of the lip) is hairy.

Dipodium pictum (Lindl.) Reichenb. f. [Figs 164 & 165]

Lowland and highland epiphyte with monopodial growth. Stems long, 1 m or more; thick, but not very tough; climbing, with numerous roots clasping substrate.

Leaves: Numerous; with long sheaths, covering stem; lanceolate-acute, more or less in one plane; slightly arching; rather stiff, but not thickened; to 40 cm long, 4.5 cm wide; bases imbricate.

Inflorescences: Axillary, erect, pale green flushed with purple; scape 14-20 cm long, rhachis 10-20 cm long, 4-10-flowered.

Bracts: Ovate-acute, to 8 mm long; pale green, flushed with purple.

Pedicel + ovary: 35 mm long, pale lilac.

Sepals: Oblanceolate-rounded, ventral side concave; to 25 mm long, 7-9 mm wide, spreading.

Petals: Narrowly obovate-apiculate; to 22 mm long, 7 mm wide, spreading to recurved. All tepals thickened; pale yellow with many, rather large, irregular red-purple spots on dorsal side, hardly visible on ventral side.

Lip: Trilobed, 22 mm long. Mid-lobe elliptic, with much deflexed sides; pale lilac; thickened; with apical, very acute, up-turned, white point, combined with short, white-hairy keel in upper third of mid-lobe. Mid-lobe 15 mm long, including point. Side lobes small, obliquely oblong-rounded, 5 mm long, 1 mm wide, pale lilac; almost touching column; apices slightly out-turned. Base of lip in contact with column; entirely taken up by wide, pale yellow, finely hairy keel.

Fig. 166. *Cymbidium roseum* J.J. Sm. Newly opened flower with yellow keels. Alt. 1600 m. September. 1.7×.

Fig. 167. *Cymbidium roseum*. 0.6×.

Column: 12 mm long, 5 mm wide; white, with bright yellow apex. Operculum white. Flowers last for two days.
Distribution: Malaysia, Java, Sumatra.

GENUS CYMBIDIUM Swartz, 1799

A medium-sized genus (about 50 species) of epiphytes and terrestrials, distributed from Madagascar to South East Asia, Japan and Australia. Many *Cymbidia* are beautiful orchids with large, long-lasting flowers, and members of the genus, including a number of hybrids, are often seen in cultivation.

The plants are pseudobulbous, but the pseudobulbs are generally hidden by the long, closely overlapping leaf-sheaths. The leaves are long and narrow. The large-flowered inflorescence arises from a leaf-axil near the base of the pseudobulb. Sepals and petals are similar, and usually spreading. The generic name is probably derived from L. 'cymba' = a boat, with reference to the shape of the lip.

Cymbidium roseum J.J. Sm. [Plate 5.2; Figs 166 & 167]
A handsome terrestrial, growing in the mountains. The pseudobulbs are small, closely spaced, ovoid, slightly flattened, 3 cm high, 2 cm wide, completely hidden by leaf-bases.
Leaves: Up to eight, 20-30 cm long, to 2.7 cm wide; oblong; unequally and obtusely bilobed at apex; rather stiff; with depressed mid-vein. Bases clasping stem, upper half of leaf spreading.
Inflorescence: Erect, axillary, 25 cm long; stalk covered with 5-10 cm long, acute, green sheaths; 2-3-flowered.
Bracts: Triangular-acuminate, 6 mm long.
Flowers: Large, 6 cm wide across sepals, long persistent, and with rather faint, pleasant fragrance.
Pedicel + ovary: 3 cm long.
Dorsal sepal: Lanceolate-apiculate, 4.5 cm long, 1.7 cm wide.
Lateral sepals: Lanceolate-apiculate, widest in the middle; slightly oblique; 4.2 cm long, 1.5 cm wide. All sepals very pale yellowish-green to white; mid-vein slightly darker green on dorsal side.
Petals: Lanceolate-apiculate, 4.2 cm long, 1.3 cm wide. All sepals and petals rather fleshy; spreading.
Lip: Trilobed, 4 cm long, white. Mid-lobe 1.3 × 1.3 cm, rounded, curved downwards; apex glabrous, rest of mid-lobe hairy, esp. centre. Centre of mid-lobe with large, ovoid marking, bright yellow when flower opens, reddish purple in older specimens. Blade of lip with two high, hairy keels, forming a V when seen from the front; bright yellow, turning reddish purple.
Column: 3 cm long; with down-curved apex; very pale green, flushed with purple; apex pale green. Operculum pale yellow.
Distribution: Malaya, Sumatra, Java.

Cymbidium finlaysonianum Lindl. [Figs 168 & 169]
Epiphytic, or lithophytic, forming large colonies on rocks near sea. Pseudobulbs small, completely covered by overlapping, fan-wise arranged leaf-sheaths. Sheath about 15 cm long.

Fig. 168. *Cymbidium finlaysonianum* Lindl. November.

Fig. 169. *Cymbidium finlaysonianum*. 1.3×.

Fig. 170. *Grammatophyllum speciosum* Bl. Cultivated. University of Malaya, Kuala Lumpur.

Leaves: About 5, long, erect, very stiff and leathery; linear-obtuse, slightly widening towards obtusely bilobed apex; lobes unequal: one up to 1 cm longer than the other; rounded, slightly falcate; shorter lobe almost truncate; leaves 70 cm long, or more, 2.5-4 cm wide; lower half V-shaped in section.
Inflorescence: Pendulous, 50-90 cm long, sparsely many-flowered (about 15 flowers). Flowers persistent (about two weeks), faintly scented.
Bracts: Very small, triangular-acute, 2.5 mm long, 3 mm wide.
Pedicel + ovary: 3.5 cm long.
Sepals: Oblanceolate-acute, 2.8-3 cm long, 0.8 cm wide; spreading; rather stiff and thickened; brownish yellow, with reddish-purple apices; flushed with purple in centre of basal part.
Petals: Elliptic-acute, not much spreading; 2.6 cm long, 0.9 cm wide; brownish yellow, with purple stripe in middle.
Lip: Trilobed, 3 cm long (when straightened). Side lobes erect, almost touching column. Mid-lobe ovate-acute, much deflexed. Base of lip with two long, narrow, purple keels, ending in yellow spot at base of mid-lobe. Side lobes and base of lip pale purple, with dark purple veins. Mid-lobe with white base and purple apex.
Column: Long, slender, curving; brownish yellow, flushed with purple. Operculum yellow. Pollinia triangular, in two groups with one large and one small pollinium each.
Distribution: Thailand, Cambodia, Vietnam, Malaysia, Philippines.

GENUS GRAMMATOPHYLLUM Blume, 1825

A small (12 species) South East Asian genus of epiphytes, a few of which are among the largest of all orchids, having very long and proportionally not much thickened pseudobulbous stems: 3 metres long, or more, by 5 cm wide. Other species are much smaller, with proportionally thicker pseudobulbs. The leaves are always long, narrow, and two-ranked.

The etymology of the generic name is not clear. The Greek word 'gramma' means 'a marking, a letter' and 'phyllon' = a leaf. Perhaps the name refers to the markings on the perianth segments.

One of the two Malayan species, *Grammatophyllum speciosum* Bl., the 'Tiger Orchid', can be seen in Figure 170. This huge lowland epiphyte is here cultivated as a terrestrial. It was not flowering when the photograph was taken. This species has long inflorescences, 2 m long or more, with *Cymbidium*-like flowers as large as 10 cm across. The flowers are pale yellow, with large orange-brown spots. The species is distributed from Thailand, Cambodia and Vietnam through Malaysia and Indonesia to the Philippines.

GENUS EULOPHIA R. Brown ex Lindley, 1823

A large terrestrial genus (more than 200 species) with pan-tropical distribution, best developed in Africa, where several, often very beautiful species are found. The six Malayan species are much less attractive.

The sympodium arises from a more or less thickened rhizome, or from a short and rather thick pseudobulb. The leaves are usually either grass-like, or wide and plicate.

Fig. 171. *Eulophia nuda* Lindl. August.

Fig. 172. *Eulophia nuda*. 2.9×.

The inflorescence is erect and many-flowered. Sepals and petals are similar; the lip is trilobed and provided with a spur (except in the Malayan *Eulophia zollingeri* (Reichenb. f.) J.J. Smith, which is also unusual in being a saprophyte). The generic name means 'handsomely crested', from Gk 'eu-' = good, and 'lophos' = a comb.

Eulophia nuda Lindl., syn. *E. squalida* Lindl. [Figs 171 & 172]

A terrestrial, common in lowland forest. Pseudobulbs small, subterraneous.

Leaves: 3 to 6, with long sheaths. Petioles grooved, up to 25 cm long. Blade narrowly elliptic-acute, plicate; up to 55 cm long, 5.5 cm wide.

Inflorescence: Sparsely many-flowered, with a few flowers open at the same time. Scape to 95 cm long, with 4 large, oblong-obtuse stem bracts, to 7 cm long, 1.8 cm wide.

Bracts: Floral bracts lanceolate-acute, 1-1.5 cm long.

Pedicel + ovary: 2.5 cm long.

Sepals: Oblong-subacuminate, 20 mm long, 5 mm wide, spreading; olive-green, flushed with purple.

Petals: Narrowly ovate-acute, with up-turned apices; not much spreading; 17-18 mm long, 8 mm wide; white with pale green base; sometimes with pale green margins in basal half and mauve venation in centre.

Lip: Trilobed; white, with dense mauve venation; base yellow and reddish brown. Side lobes nearly semi-circular, 5 mm long, 9 mm wide, erect. Mid-lobe deflexed, triangular, with very wavy margins; 7 mm long, 8 mm wide; central spot with yellow-brown venation. Blade with 6 low keels, yellow with minute purple spots; one additional keel on base of each side lobe. Base of lip prolonged into short spur, 5 mm long from base of lateral sepals.

Spur: Olive-green, flushed with purple.

Column: Winged, white or pale green, ventral side densely spotted with mauve, base brown. Column-foot long, bifid. Anther pale yellow with mauve margins, and two small lobes on top.

Capsules: Fusiform, conspicuously 6-ridged; 5.5 cm long, 1 cm wide.

Distribution: Nepal, India, Sri Lanka, China, Burma, Thailand, Laos, Cambodia, Vietnam, Malaysia, Indonesia, New Guinea, Philippines.

GENUS BROMHEADIA Lindley, 1841

A small South-East Asian-Australian genus with some 17 species, 7 of which are found in Malaya. The flowers in distichous inflorescences are very similar in all species, being white or pale yellow, with purple and yellow markings. *Bromheadia*-flowers are rather large, pretty, and often nicely scented, but all are very briefly open. In some species, they are open for a few hours in the morning only. The flowering is more or less pronouncedly gregarious. The vegetative parts can differ considerably. One group of Malayan species, including the common terrestrial *Bromheadia finlaysoniana*, has 'normal' leaves, with blades spreading horizontally. The other group is vegetatively similar to *Oberonia* or to the Aporum or Oxystophyllum sections of *Dendrobium*: the plants are laterally flattened, and have thickened, pointed leaves which are all in one plane.

Fig. 173. *Bromheadia finlaysoniana* (Lindl.) Miq.

Fig. 174. *Bromheadia alticola* Ridl. December. 0.7×.

Fig. 175. *Bromheadia alticola*. 3×.

Fig. 177. *Bromheadia brevifolia*. 2.7×.

Fig. 176. *Bromheadia brevifolia* Ridl. Alt. 1100 m. December. 1.9×.

Bromheadia finlaysoniana (Lindl.) Miq. [Fig. 173]

A common lowland terrestrial, found in all parts of Malaya, often on road-sides with open scrub. Stems 1 m high, or more, leafy in lower part only.

Leaves: Two-ranked, fleshy, tapering towards bilobed apex; 12 cm long, 3 cm wide, with sheaths 4 cm long.

Inflorescence: Terminal, 10 cm long or more; many-flowered, only one to three flowers appearing at the same time.

Bracts: Two-ranked, alternate, closely spaced, stiff, cymbiform; to about 1 cm long.

Pedicel + ovary: 1-1.5 cm long.

Sepals: Lanceolate-acute, 2.5-3.8 cm long, 0.6-1 cm wide; spreading, white. Lateral sepals falcate.

Petals: Falcately elliptic-acute, spreading, white; 1.2 cm wide, slightly shorter than sepals.

Lip: Trilobed, 3.3 cm long. Side lobes erect, close to column; free ends slightly longer than column, pointed. Mid-lobe about 1 cm long; large, central, warty, yellow callus. Margins white, often with purple venation.

Column: Long, slender, arching, white.

Distribution: Laos, Cambodia, Vietnam, Thailand, Malaysia, Indonesia.

Bromheadia alticola Ridl. [Figs 174 & 175]

Lowland and highland epiphyte with tufted stems, 25-60 cm long, leafy in upper part only. Leafless part about 10 cm long, covered with sheaths.

Leaves: Straight, stiff, but not much thickened, 10-20 cm long, 1.5-2 cm wide; oblong, unequally bilobed at apex, lobes rounded. Leaf-sheaths 3.5 cm long.

Inflorescence: From top of stem, stalk very short; rhachis 6-10 cm long, with many, persistent, closely two-ranked bracts which are laterally flattened, conduplicate, more or less rhomboid, and 6-10 mm long and wide. Only one flower present at a time, very shortly persistent. Flowering gregarious. Pedicel + ovary 1 cm long.

Sepals: Oblanceolate-obtuse, 19-25 mm long, 6-8 mm wide, pale yellow. Laterals slightly falcate, 7-9 mm wide.

Petals: Oblanceolate-acute, 19-25 mm long, 5-8 mm wide, pale yellow.

Lip: 16-18 mm long, 10 mm wide across side lobes. Side lobes short, with long base; erect; pointed at distal end; white. Mid-lobe triangular, straight, 5 mm long, 5 mm wide at base; papillose, bright yellow. Base of lip yellow, with 10 mm long keel, slightly raised and rounded in distal end, and near column.

Column: 12-14 mm long, straight, 2-winged, pale yellow. Operculum white.

Distribution: Malaya, Thailand.

Bromheadia brevifolia Ridl. [Figs 176 & 177]

A highland epiphyte. Stems tufted, often leafy throughout; 10-25 cm long, 1 cm wide across leaves.

Leaves: Numerous, laterally flattened; closely two-ranked, almost touching; lanceolate-acute; much thickened, stiff; pointing upwards at 45° from stem; to 10 mm long, 3 mm wide; internodes about 5 mm. Base of leaf and leaf-sheath on young stems sometimes purple.

Pedicel + ovary: 14 mm long; green, flushed with purple. Buds deep purple, turning pale yellow before opening. Flowers very shortly persistent: opening in the morning, they wilt before noon.

Fig. 178. *Bromheadia truncata* Seidenfaden. November. 1.1×.

Fig. 179. *Bromheadia truncata*. 3×.

Fig. 180. *Bromheadia scirpioidea* Ridl. September. 1.5×.

Fig. 181. *Bromheadia scirpioidea*. 2.5×.

Sepals: Oblong-obtuse, 11-14 mm long, 3.5 mm wide; pale yellow, with irregular purple streaks, especially in centre and near base.

Petals: Oblong-obtuse, 11 mm long, 3 mm wide, pale yellow.

Lip: 11 mm long, pale yellow, trilobed. Mid-lobe straight; margins wavy, with incision in front; purple band, 1.5 mm wide, across base; yellow spot, 2 mm wide, at apex, with yellow keel continuing into the purple band. Margins white at apex. Side lobes white, pointed in distal part; slightly spreading, short. Base of lip with yellow keel, getting slightly wider, and papillose, towards column; with small, purple spots near base.

Column: Pale green, purple near lip attachment.

Anther: Pale yellow.

Bromheadia truncata Seidenfaden [Figs 178 & 179]

Lowland and highland epiphyte. Stems tufted, to 10 cm long, laterally compressed.

Leaves: About 7, longest in upperpart of stem, very small at base of stem; laterally compressed, thickened, very stiff; oblong to linear, sharply pointed; often slightly arching; to 80 mm long, 4-5 mm wide.

Inflorescences: From axil of small leaf at apex of stem; arising among numerous chaffy bracts; with only one, or rarely two, flowers open at a time. Flowering strictly gregarious. Flowers very shortly persistent (closed before noon); fragrant.

Pedicel + ovary: 16 mm long.

Sepals and petals: Lanceolate-acute, 16-18 mm long, 4 mm wide, pale yellow.

Lip: 18 mm long, indistinctly trilobed. Side lobes encircling column; pale yellow. Mid-lobe 7 mm long, 5 mm wide; rounded, with very wavy margins; margins white; centre with elliptic, yellow marking and warty callus. Purple band at base of mid-lobe. Base of lip pale yellow, with yellow keel, 9 mm long, 1.3 mm wide; keel with raised, rounded, red-hairy end, pointing towards apex of lip; basal part of keel with 2 or 3 longitudinal rows of irregular, purple spots.

Column: 12 mm long, winged, very pale yellow; base spotted with purple.

Anther: White.

Distribution: Thailand, Malaya, Sumatra.

Bromheadia scirpioidea Ridl. [Plate 5.3; Figs 180 & 181]

Lowland epiphyte with densely tufted stems, up to 15 cm long, sheathed at base. Internodes 12 mm long.

Leaves: 2-3 on each side of stem, laterally flattened, very acute, narrow, stiff, sometimes almost terete; to 9 cm long, 1.5-3 mm wide; forming narrow angle with stem; arching. Base of leaf clasping stem.

Inflorescence: Terminal, very short, many-flowered; only one or two flowers open at the same time, and for a few hours only.

Bracts: Numerous; short, acute, thin and scarious.

Pedicel + ovary: 7 mm long.

Sepals: Oblong-acute, 11-15 mm long, 3-4 mm wide; not much spreading; pale yellow or pale green.

Petals: Oblong-acute, slightly shorter than the sepals; to 4 mm wide; not much spreading; pale yellow or pale green.

Lip: Trilobed, 12-14 mm long, 5 mm wide across mid-lobe. Side lobes wide; truncate in front; incurved over column; pale green. Mid-lobe quadrangular-retuse, 3-4 mm long, with wavy white margins and large, rhomboid, papillose yellow callus; apex

Fig. 182. *Polystachya concreta* (Jacquin) Garay & Sweet. Alt. 1350 m. October. 0.7×.

Fig. 183. *Polystachya concreta*. 10.2×.

Fig. 184. *Octarrhena parvula* Thw. Alt. 1500 m. December. 6.4×.

deflexed. Blade white, with two large purple patches near mid-lobe, and yellow-green keel between patches and base. Keel higher in apical part; sometimes spotted with purple near base.

Column: Long, slender, slightly curved; white or pale green, with a few purple spots along base, near attachment of lip. Column-wings rounded, slightly higher than anther. Operculum white or cream-coloured. Rostellum deeply bifid.

Distribution: Malaya, Borneo.

GENUS POLYSTACHYA Hooker, 1825

A large epiphytic genus (about 200 species), with pan-tropical distribution. Most of the species are African, and there is only one representative in Malaya. The plants have pseudobulbs and numerous leaves. The inflorescence arises at the apex of the pseudobulb and is generally erect and many-flowered.

Polystachya-flowers are usually small, but easy to recognize: they are not resupinate, and the lateral sepals form a distinct, hood-like mentum which is then found at the top of the flower. Etymology: 'polys' = many; 'stachys' = an ear of corn, both words are Greek.

Polystachya concreta (Jacquin) Garay & Sweet, syn. *Polystachya flavescens* (Bl.) J.J. Sm. (and many other synonyms) [Figs 182 & 183]

Lowland and mountain epiphyte. Pseudobulbs small, covered by leaf-bases.

Leaves: 2-3, at base of stem; oblanceolate, unequally and conniventl bilobed at apex; 10 cm long, 1.6 cm wide, with long sheaths. Lowest leaf small, 2.5 cm long, narrowly elliptic. Scape to 20 cm long, covered with long, acute, scarious sheaths. Rhachis about 5 cm long, often branched, or with arrested branches; with sparse, short, white hairs.

Inflorescence: Sparsely many-flowered.

Bracts: Triangular-very acute, 4 mm long, glabrous.

Pedicel + ovary: 7 mm long, glabrous.

Dorsal sepal: Triangular-acute, 4 mm long, 1.5-2 mm wide; not much spreading, forming hood with petals; pale green, margins slightly flushed with mauve.

Lateral sepals: Obliquely triangular-acute, 4.5 mm long; with wide base, forming mentum, 3.5 mm long; pale green, with slight mauve flush on margins. Apex of mentum mauve.

Petals: Linear-apiculate, 3 mm long, 0.6 mm wide; very pale green; apex very pale mauve.

Lip: Trilobed, 4 mm long, 3 mm wide across side lobes. Mid-lobe more or less rounded; recurved, especially at pointed apex; 2 × 2 mm; cream-coloured, with short, dense white hairs in central and basal part. Side lobes oblong-rounded, falcate, close to base of mid-lobe; white, with very faint flush of mauve. Base of lip white, covered with dense, short, white hairs; with 2 mm long, red callus, more or less covered with hairs.

Column: 1.5 mm long, pale green. Column-foot 3 mm long, pale green with mauve apex. Operculum yellow.

Distribution: Pantropical.

Fig. 185. *Octarrhena condensata* (Ridl.) Holttum. Alt. 1650 m. June. 0.7×.

Fig. 186. *Octarrhena condensata*. 4.5×.

Fig. 187. *Thelasis macrobulbon* Ridl. Alt. 1050 m. December. 1.3×.

Fig. 188. *Thelasis macrobulbon*. 0.5×.

Fig. 189. *Thelasis macrobulbon*. 5.7×.

GENUS OCTARRHENA Thwaites, 1864

A small South Asian genus with two representatives in Malaya. Both are small, inconspicuous epiphytes. The leaves are thickened and laterally compressed, and rather similar to *Oberonia*-leaves. The inflorescence, too, resembles that of an *Oberonia*, having numerous, very small flowers. However in *Octarrhena*, the inflorescence is lateral, not terminal as in *Oberonia*. The 8 pollinia explain the generic name, composed of the Greek words 'okto' = eight, and 'arren' = a male.

Octarrhena parvula Thw. [Fig. 184]

Small highland epiphyte. Stem flattened, 3-6 cm long, branching at base only; covered with leaf-sheaths.

Leaves: Thickened, laterally flattened; conspicuously jointed to sheath; to 25 mm long, 2.5 mm wide, very acute. Internodes 4.5 mm long.

Inflorescence: Axillary, to 25 mm long; sparsely many-flowered, only a few flowers open at a time.

Bracts: 1.5 mm long, acute.

Pedicel + ovary: 1.5 mm long.

Sepals: Triangular-obtuse, much spreading; 1.5 mm long.

Petals: Rounded, 0.5 mm long, spreading.

Lip: Ovate, rounded, 0.5 mm long.

Column: Short, straight. All tepals and column pale green.

Distribution: Sri Lanka, Malaya, Sumatra, Java.

Octarrhena condensata (Ridl.) Holttum [Figs 185 & 186]

Highland epiphyte. Stem 10-20 cm long, leafy only in upper part.

Leaves: To 60 mm long, 4.5-8 mm wide; laterally compressed; base imbricate, apex acute.

Inflorescence: Lateral, about 40 mm long; many-flowered, with most flowers open at the same time, starting from the apex.

Bracts: Caudate, 2 mm long.

Flowers: Yellow, turning orange when old.

Sepals: Triangular-acute, 1.5 mm long.

Petals: Narrowly triangular-acute, 1 mm long.

Lip: Narrowly triangular-acute, 1.5 mm long; with up-turned apex.

Anther: White.

GENUS THELASIS Blume, 1825

A genus of rather small epiphytes, either pseudobulbous with one or two terminal leaves, or without a pseudobulb and having several, laterally flattened leaves. The inflorescence is axillary, with many, very small and closely placed green or yellow flowers. Sepals and petals are similar; not much spreading, but with recurved apices. The lip is wide at the base, and narrows towards the apex. The column has no foot. Five species are recorded for Malaya.

Thelasis macrobulbon Ridl. [Figs 187, 188 & 189]

A highland epiphyte. Pseudobulbs very closely spaced, broadly ovate, 2.5 × 2.5 cm;

Fig. 190. *Phreatia listrophora* Ridl. Alt. 1500 m. June. 1×.

Fig. 191. *Phreatia listrophora*. 8×.

Fig. 192. *Phreatia linearis* Ridl. Alt. 1000 m. December.

Fig. 193. *Phreatia linearis*. 6.4×.

Plate 5
5.1. *Ceratostylis ampullacea* Kränzl. 4.8×.
5.2. *Cymbidium roseum* J.J. Sm. 1.2×.
5.3. *Bromheadia scirpioidea* Ridl. 3.1×.
5.4. *Gastrochilus calceolaris* Don. 7.3×.
5.5. *Uncifera tenuicaulis* (Hook. f.) Holttum. 7.3×.

5.1	5.2	5.3
5.4	5.5	

slightly laterally flattened; two-leaved, one leaf much larger than the other.

Leaves: Large leaf oblong-acute, widening towards apex; thickened; about 20 cm long, 2.5-3.5 cm wide. Small leaf with sheath completely covering pseudobulb; blade ovate-acute, 3-4 cm long, 1.2-1.5 cm wide.

Inflorescence: From base of pseudobulb, scape 12-20 cm long, with four acute sheaths, 0.6-1.3 cm long (upper ones smaller). Rhachis 8-12 cm long, many-flowered, nearly secund; about ⅓ of the flowers open at the same time.

Bracts: Triangular-acuminate, 5 mm long, dark brown.

Flowers: Green, turning yellow when wilting.

Pedicel + ovary: 2-3 mm long, thickened.

Sepals: Ovate-rounded, not much spreading; 3 mm long, 2 mm wide.

Petals: Elliptic-rounded, 2.5 mm long, 1.5 mm wide, apex slightly spreading.

Lip: With triangular-obtuse blade from narrow base; blade 2 mm long, 1.5 mm wide; apex decurved.

Column: Green, with two long processes, the apices of which touch above the anther.

GENUS PHREATIA Lindley, 1830

Vegetatively, this genus is rather similar to *Thelasis*. Both genera are many-flowered, but Phreatia less densely so. The small flowers have a distinct column-foot and a mentum, and the lip is typically narrowed at the base. Eight species out of a total of about 120 are found in Malaya. The genus as such is distributed in South Asia and Australia, and its name is probably derived from the Greek 'phrear' = a well.

Phreatia listrophora Ridl. [Figs 190 & 191]

Small highland epiphyte with tufted stems, and pseudobulbs 4 mm long, 5 mm wide. The pseudobulbs are single- or two-leaved, one leaf being much smaller than the other.

Leaves: Oblong-acute, 4.5-14 cm long, about 1 cm wide.

Inflorescence: From base of pseudobulb, many-flowered. Scape stiff, erect, very slender, 6-11 cm long.

Bracts: Narrowly triangular-acute; scarious, persistent; up to 5 mm long.

Flowers: Not resupinate; white.

Dorsal sepal: Narrowly ovate-acute, recurved; 1-2 mm long, to 1 mm wide.

Lateral sepals: Triangular-acute, 2 mm long and wide.

Petals: Ovate-acute, smaller than the dorsal sepal.

Lip: Arrow-like, with long, slender, up-turned claw, suddenly widening to triangular-acute blade, 3 mm long.

Anther: Cream-coloured.

Distribution: Malaya, Thailand.

Phreatia linearis Ridl. [Figs 192 & 193]

Small highland epiphyte. Pseudobulbs well-spaced on rhizome; small, conical, 3-5 mm high and wide, 1-2-leaved; one leaf much smaller than the other.

Leaves: Narrowly lanceolate-acute from narrow base, 4-10 cm long (larger leaf), 0.6-0.8 cm wide.

Inflorescence: From base of pseudobulb, with 4-5 lanceolate-acute, clasping sheaths; many-flowered, slender.

Fig. 194. *Phreatia densiflora* (Bl.) Lindl. Alt. 1500 m. December. 0.4×.

Fig. 196. *Phreatia pusilla* (Bl.) Lindl. Alt. 1400 m. March.

Fig. 195. *Phreatia densiflora*. 3.4×.

Bracts: Lanceolate-acute, 3 mm long, 1 mm wide.
Flowers: White, very small.
Dorsal sepal: Ovate-subacute, 1.2-2 mm long, 1 mm wide.
Lateral sepals: Triangular-acute, 2 mm long, 1.5 mm wide.
Petals: Ovate-subacute, 1.5 mm long, 0.7 mm wide.
Lip: 1.5 mm long, 1 mm wide; V-shaped, with blunt sac at base.
Distribution: Probably endemic.

Phreatia densiflora (Bl.) Lindl., syn. *Rhyncophreatia densiflora* (Bl.) N. Hallé ined. [Figs 194 & 195]

Highland epiphyte with short stem, completely covered by leaf-bases.
Leaves: 10-12, linear to oblong; unequally bilobed at apex, longer lobe rounded, oblique; shorter lobe acute, connivent. Whole leaf to 30 cm long, 1.8 cm wide; basal part conduplicate. Leaves fan-wise arranged in one plane, with imbricate bases.
Inflorescence: Axillary; scape to 20 cm long, with a few acute sheaths, to 1.5 cm long. Rhachis to 10 cm long, spicate, rather densely many-flowered; flowers start opening at base of inflorescence.
Bracts: Floral bracts dark brown, acute, 2 mm long.
Pedicel + ovary: 3 mm long.
Dorsal sepal: Triangular-rounded, to 1.2 mm long, 0.9 mm wide.
Lateral sepals: Triangular, 1 mm long, 1 mm wide at base; with wide base, forming mentum.
Petals: Ovate-rounded, 1 mm long. Lip attached to column-foot; widening from narrow base; rounded apex sharply deflexed; 1.2 mm long, 1 mm wide. All tepals white.
Column: Less than 1 mm long, pale green; column-foot about 1 mm long.
Anther: With brown spot (darker in mature flowers).
Distribution: Malaysia, Thailand, Indonesia, Philippines.

Phreatia pusilla (Bl.) Lindl. [Fig. 196]

Small highland epiphyte. Stem very short, with leaves arising in a fan-like tuft.
Leaves: Oblong or oblanceolate; unequally and obtusely bilobed at apex, imbricate at base; 3-15 cm long, up to 0.6 cm wide.
Inflorescence: Many-flowered, 7-15 cm long.
Bracts: Lanceolate-acute, 2-3 mm long.
Sepals: Triangular-obtuse, 1 mm long, white; laterals wider than dorsal sepal.
Petals: Very small, suborbicular, white.
Lip: Oblanceolate-obtuse, apex deflexed; 1.5 mm long, white.
Column: Nearly as long as dorsal sepal; straight, pale green.
Anther: White.
Distribution: Malaya, Java, Sumatra.

Phreatia laxiflora (Bl.) Lindl. [Figs 197 & 198]

Highland epiphyte with short, flattened stem, covered with imbricate leaf-bases. Stem to 2.4 cm wide across leaf-bases.
Leaves: 5-8, very narrowly lanceolate; unequally (slightly) and obtusely bilobed at apex; rather stiff and fleshy; to 11 cm long (without sheath), 1.2 cm wide. Sheaths 2-3 cm long from joint with leaf. Mid-vein depressed, forming keel on lower side; keel terminated at apex by small, up-turned point.

Fig. 197. *Phreatia laxiflora* (Bl.) Lindl. Alt. 1340 m. November. 0.6×.

Fig. 198. *Phreatia laxiflora*. 6.4×.

Fig. 199. *Taeniophyllum tjibodasanum* J.J. Sm. Alt. 1200 m. October. 2.7×.

Fig. 200. *Taeniophyllum tjibodasanum*. 12×.

Inflorescence: Lateral, from axil of one of the lower leaves; scape 9-11 cm long, erect; with 4 stem bracts, 1.2 cm long. Rhachis to 11 cm long, arching, many-flowered. Flowers start opening in the middle of the rhachis. Buds conspicuously yellow.
Bracts: Lanceolate-acute, more or less scarious, 5 mm long, 1 mm wide.
Flowers: Very small, facing apex of rhachis; white; faintly and rather unpleasantly scented.
Pedicel + ovary: To 2 mm long, green.
Dorsal sepal: Triangular-obtuse, 1.5-2 mm long, 1 mm wide.
Lateral sepals: Obliquely triangular-acute, 1.6 mm long; with wide base. Bases not joined, but forming short (1 mm) mentum.
Petals: Lanceolate-acute, 1 mm long.
Lip: Joined with narrow claw to column-foot; blade triangular-acute, abruptly widening from narrow base; margins up-turned; 1.2 mm long, 0.9 mm wide; with fine white hairs. Rear end of blade narrowly saccate; sac 0.5 mm long.
Column: Pale green; operculum white.
Distribution: Malaya, Java, Sumatra.

GENUS TAENIOPHYLLUM Blume, 1825

This genus, and the following ones, all belong to the same large group of orchids, called the Vanda-Arachnis-tribe. They all have in common the monopodial growth, i.e. the top of the stem can grow indefinitely. Usually, the stem is provided with roots all along. Pseudobulbs are totally lacking in this group. In Africa and Madagascar, there is a similar, but smaller, group of monopodial orchids, the Angraecoids.

Taeniophyllum is thus capable of unlimited apical growth, but it is really hard to find smaller and more inconspicuous orchids than these little epiphytes. They are in fact very easily overlooked, as they are always very small and leafless. Usually, one notices them only when looking closely at something else on a branch! As *Taeniophylla* (from L. 'taenia' = a ribbon, + Gk 'phyllon' = a leaf) are leafless, the roots have taken over the assimilating function; consequently they are numerous and provided with chlorophyll. The inflorescence arises, as always in this tribe, from the leaf-axils. On close inspection of the plant, it is usually possible to detect rudimentary leaves, in the form of very small, brown scales. The flowers are always quite small in the genus.

Taeniophyllum is quite a large genus, with about 170 species, distributed in South Asia, Japan and Australia. One species only has been recorded for Africa (where similar, leafless epiphytic orchids normally belong to a different genus, *Microcoelia*). From Malaya, 17 species of *Taeniophyllum* are known.

Taeniophyllum tjibodasanum J.J. Sm. [Figs 199 & 200]
Very small, apparently leafless highland epiphyte, with numerous, pale green, rounded roots, slightly flattened on lower side.
Inflorescence: Short, 1.5 cm long; rhachis 1 cm long.
Bracts: Two-ranked, alternate; triangular-acute, 1.7 mm long.
Pedicel + ovary: 2 mm long. Only 1-2 flowers open at a time.
Tepals: All similar; lanceolate-acute, pale yellowish-green, 4 mm long, 1 mm wide; united for 0.5 mm at the base.

Fig. 201. *Taeniophyllum obtusum* Bl. February. 4×.

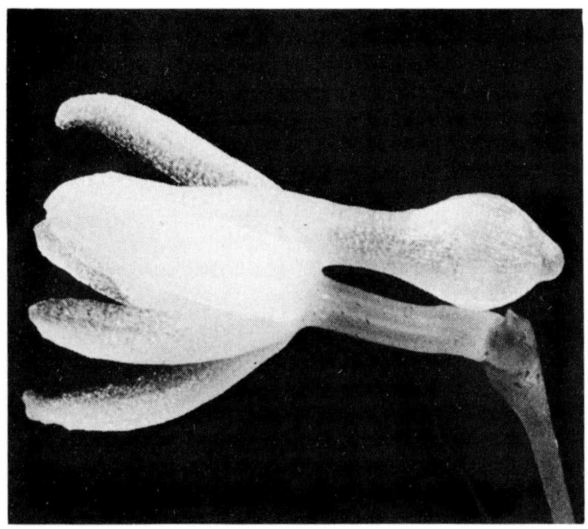

Fig. 203. *Taeniophyllum filiforme*. 11.1×.

Fig. 202. *Taeniophyllum filiforme* J.J. Sm. October. 2.2×.

Lip: Slightly shorter, narrower and more acute than sepals; with short, blunt, sac-like spur; whole lip 4.5 mm long; spur 1 mm long. Apex of lip slightly swollen, and slightly deflexed; blade with thread-like process, directed backwards.
Column: Green; operculum yellow.
Distribution: Malaya, Java.

Taeniophyllum obtusum Bl. [Fig. 201]
Very small, apparently leafless epiphyte, growing in the lowlands, often in exposed situations, near the sea. Stem very short. Roots numerous, green, slightly flattened.
Inflorescences: Very short, up to 20 mm long; zig-zag, many-flowered, only one flower open at a time.
Bracts: Triangular-acute, 1 mm long.
Flowers: Yellow.
Tepals: Lanceolate-acute, 3 mm long.
Lip: 3.5 mm long (including blunt spur), apex white, with purple spots.
Distribution: Cambodia, Thailand, Malaysia, Indonesia.

Taeniophyllum filiforme J.J. Sm. [Figs 202 & 203]
Very small, apparently leafless lowland epiphyte. Roots green, slightly flattened; with white streaks, 1 mm long, often in two parallel rows, on lower side of roots. Roots often free from substratum; numerous, thin, to 1 mm wide. Stem short, about 10 mm long, with a few, very small, acute, scarious scales.
Inflorescences: On long (to 50 mm), very slender (0.3 mm wide) stalks; with one minute, clasping stem bract with deflexed apex; many-flowered, only one flower appearing at a time.
Bracts: Floral bracts triangular-acute, 0.5-1 mm long.
Flowers: Greenish white, not much opening; not resupinate.
Pedicel + ovary: 3 mm long. Length of flower 8 mm (from apex of dorsal sepal to apex of spur).
Tepals: Oblong-obtuse, 4 mm long, 1.4 mm wide.
Lip: 6.5 mm long, with large, saccate spur. Blade triangular-cymbiform, thickened; margins much up-turned; base joined with column; 3 mm long.
Spur: 4 mm long; tubular near the base, much swollen (1.5 mm wide) towards apex; ending with obtuse point; pale green.
Column: Short, pale green. Rostellum elongated, oblong-subtruncate; white. Operculum white.
Distribution: Andaman Isl., Thailand, Malaya, Indonesia.

GENUS MICROSACCUS Blume, 1825

A small, South Asian genus with five Malayan species. The plants are all small, and rooting at the base only, which is unusual in monopodials. The leaves are fleshy and laterally compressed, making the plants rather similar to species of the Aporum section of *Dendrobium*. However the *Microsaccus*-inflorescence is quite distinct, always bearing a pair of small, white, spurred flowers. Etymology: Gk 'mikros' = small; L. 'saccus' = a sac.

Microsaccus sumatranus J.J. Sm. [Figs 204 & 205]
A small highland epiphyte. Stems branching near base, tufted, to 14 cm long, often

Fig. 205. *Microsaccus sumatranus*. 9.3×.

Fig. 204. *Microsaccus sumatranus* J.J. Sm. Alt. 1500 m. December. 1.6×.

Fig. 206. *Adenoncos vesiculosa* Carr. Alt. 1340 m. September. 4.4×.

leafy in upper part only; leafless part covered with withered leaf-sheaths.

Leaves: Numerous, fleshy, laterally compressed; strictly two-ranked; lanceolate-acute, with well-marked groove on upper surface; to 11 mm long, 3 mm wide; closely placed, touching. Width of plant across leaves up to 20 mm.

Bracts: Triangular-acute, white-hyaline; as long as ovary.

Flowers: 6 mm wide, axillary; in pairs, white.

Sepals and petals: Spreading.

Dorsal sepal: Oblanceolate, rounded or obtuse; with thickened keel in apical part, on dorsal side; 3 mm long, 1 mm wide.

Lateral sepals: Oblong-obtuse, apex thickened, keeled on dorsal side; 4 mm long, 1 mm wide; slightly falcate.

Petals: Oblanceolate-obtuse, apex thickened and keeled on dorsal side; 2 mm long, 1.5 mm wide.

Lip: Rounded, with erect margins; diffuse green spot on blade.

Spur: Slightly flattened; more or less truncate at apex; 2-2.5 mm long, 1.5 mm wide.

Column: Short, white. Operculum white.

Distribution: Malaya, Sumatra.

GENUS ADENONCOS Blume, 1825

The four Malayan species are small epiphytes, vegetatively rather similar to *Microsaccus*. The leaves in *Adenoncos*, however, are not laterally flattened; they are very thick, but have a distinct groove on the upper side. The inflorescence is very short, with a few, greenish flowers which are not spurred, but provided with a concave lip-base and a papillose callus (Gk 'aden' = a gland; 'onkos' = a tumescence).

Adenoncos vesiculosa Carr [Fig. 206]

Small highland epiphyte. Stems 5-15 cm long, leafy in upper part. Internodes 3-4 mm.

Leaves: Two-ranked, much thickened; narrowly lanceolate-acute, almost terete, but with a groove on the upper side of the leaf; to 13 mm long, 3 mm wide; slightly arching.

Inflorescence: Axillary; 1-2-flowered.

Bracts: 1.5 mm long.

Flowers: Small, inconspicuous.

Pedicel + ovary: 2.5 mm long.

Sepals: Lanceolate-obtuse, 2.5 mm long, 1.2 mm wide.

Petals: Almost linear; apiculate (or subacute), 2.5 mm long, 0.5 mm wide.

Lip: 3 mm long, 2.8 mm wide; widening from narrower base; thickened; firmly attached to column-foot. Blade concave, with low papillose keel in basal part. Apex rounded, but with small point near apex, on lower side.

Column: Short, thick, pale green.

Anther: Brown. Rostellum shortly pointed.

Capsules: 7 mm long.

Distribution: Thailand, Vietnam, Malaya.

Fig. 207. *Thrixspermum scopa* (Hook. f.) Holttum. July. 0.7×.

Fig. 208. *Thrixspermum scopa*. 3.3×.

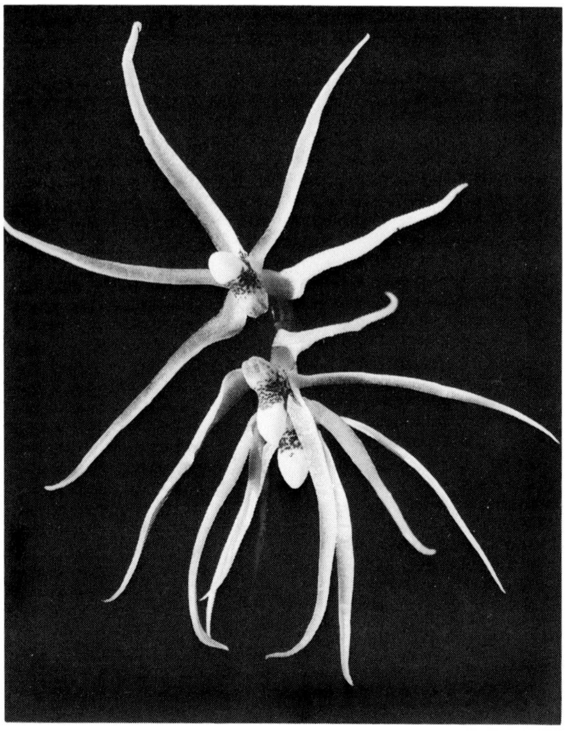

Fig. 209. *Thrixspermum arachnites* (Bl.) Reichenb. f. Alt. 800 m. September. 1.2×.

Fig. 210. *Thrixspermum arachnites*. 3.6×.

GENUS THRIXSPERMUM Loureiro, 1790

Orchids in this genus have a distinct column-foot, to which the lip is firmly joined. The lip is saccate, with a callus on the anterior wall of the sac. At least two sections are generally recognized in this genus: one section (Thrixspermum or, previously, Orcidice) has laterally compressed bracts and two-ranked flowers; sepals and petals are often long and slender. The other section (Dendrocolla) has flowers with short, broad sepals and petals, and flowers facing in all directions. Some Thrixspermum-flowers are rather large, pleasantly scented and quite attractive, but all are very shortly persistent. A number of species have gregarious flowering. The seed capsules are very long and slender. The Malayan species number about 28. Etymology: Gk 'thrix' = hair; 'sperma' = a seed.

Thrixspermum scopa (Hook. f.) Holttum [Figs 207 & 208]

A lowland epiphyte with long, climbing stems and numerous aerial roots; internodes 1-5 cm long.

Leaves: Narrowly elliptic or oblanceolate, unequally and obtusely bilobed at apex; to 12 cm long, 2.5 cm wide.

Inflorescences: One to many, many-flowered, only 1-3 flowers open at a time. Flowering gregarious. Scape 2-20 cm long; stiff, erect, with one amplexicaul stem bract. Rhachis 2-10 cm long, up to 1 cm wide.

Bracts: Laterally compressed, falcate, blunt; 8 mm long. Scape and rhachis green, with purple spots. Margins of bracts purple.

Pedicel + ovary: To 10 mm long.

Tepals: All cream-coloured.

Sepals: Very narrowly lanceolate-acute; spreading; 35-70 mm long, 3 mm wide.

Petals: Sublinear-acute, 30-65 mm long, 1.5-2 mm wide.

Lip: Saccate, trilobed, 10 mm long. Mid-lobe cylindrical-obtuse, very fleshy, pure white; 6 mm long, 2.5 mm wide. Sac 3.5 mm long (from base of mid-lobe). Side lobes convergent above column; falcate, projected forwards, 2 mm long, 4 mm wide; apex white, base very pale mauve with brown spots; very finely hairy. Base of lip very pale mauve, with reddish brown spots and conspicuous, ligulate callus, 1 mm wide, between side lobes.

Column: Short, white; foot spotted with purple. Operculum pale yellowish green.

Seed capsules: Sharply ridged; to 10 cm long, 1 cm wide.

Distribution: Thailand, Malaysia.

Thrixspermum arachnites (Bl.) Reichenb. f. [Figs 209 & 210]

Lowland epiphyte. Stem to 15 cm long, slightly flattened. Internodes 1-1.5 cm long.

Leaves: Oblanceolate or oblong, slightly falcate, thick and stiff; equally or unequally bilobed at apex, obtuse; to 12 cm long, 2.5 cm wide.

Inflorescences: To 14 cm long. Flowering gregarious, with 1-3 flowers open at a time on each inflorescence. Flowers very shortly persistent: opening in the morning, they are wilted at noon; fragrant, with *Convallaria majalis*-like scent. Rhachis 10-45 mm long.

Bracts: Obliquely triangular-acute, bases imbricate; laterally compressed; 7 mm long.

Pedicel + ovary: 10 mm long; scurfy.

Tepals: All similar, pale yellow, very narrowly lanceolate-acute; 35 mm long, 3 mm

Fig. 211. *Thrixspermum calceolus* (Lindl.) Reichenb. f. November. 0.5×.

Fig. 212. *Thrixspermum calceolus*.

wide; spreading. Petals slightly narrower than sepals. Lateral sepals with slight bend in vertical plane, near base, which is 4 mm wide.

Lip: Trilobed, 10 mm long. Mid-lobe saccate, conical, very fleshy; white at apex, with brown spots or reticulate pattern around base; 6 mm long; base finely papillose. Base of lip saccate; sac slightly bilobed at apex; 3 mm long; pale brown, with brown spots in basal part. Side lobes almost rectangular; erect, hiding column; 2 mm high, 5 mm long; white, with purple spots; front edge with irregular teeth. Blade of lip white, with purple spots and thin, ligulate callus, curving forwards.

Column: Very short, white. Operculum white.

Distribution: Malaysia, Indonesia.

Thrixspermum duplocallosum Holttum

A highland epiphyte. Stem 6 cm long, winding.

Leaves: Slightly fleshy, narrowly elliptic to oblong; more or less unequally and conniventy bilobed at apex; lobes rounded; 6.5 cm long, 1.3 cm wide.

Inflorescences: Numerous; very slender, many-flowered (only one flower open at a time); to 10 cm long. Scape to 8 cm long, 0.5 mm wide, with one small, acute stem bract. Rhachis thickened, to 3 cm long, with many closely placed, imbricate, fleshy bracts.

Bracts: 2 mm long; green, often flushed with purple.

Pedicel + ovary: 5 mm long, slightly scarious.

Dorsal sepal: Obovate-rounded, 7.5 mm long, 3.5 mm wide.

Lateral sepals: Oblanceolate-rounded, 6 mm long, 3 mm wide.

Petals: Slightly oblique; obovate-apiculate (apiculus very small, mauve); 6 mm long, 4 mm wide. Sepals and petals white.

Lip: 5 mm long, trilobed. Side lobes triangular-rounded, erect, 3 mm long; orange-yellow; posterior margin white; apex irregular. Mid-lobe triangular, with wavy margin; deflexed; 2 mm long; orange-yellow. Base of lip forming shallow sac, 3 mm long; conspicuous, erect, bifid yellow keel in front of sac-opening; apices of keel decurved. Floor of sac crimson, with large, narrow, longitudinal, elliptic callus, 2 mm long; with a blunt point facing lip.

Column: Short, 1.5 mm long, white.

Seed capsules: Slender, grooved, 25 mm long, 2 mm wide.

Distribution: Endemic to Malaya.

Thrixspermum calceolus (Lindl.) Reichenb. f. [Figs 211 & 212]

This large monopodial is epiphytic, or lithophytic on rocks near the sea. The branching and stiff stems can be several metres long. They are creeping and winding, leafy throughout, and have numerous, long and sturdy aerial roots. The leaves are alternate, twisted at the base and more or less in one plane. They are up to 7.5 cm long, 2.5 cm wide, stiff and thickened, narrowly elliptic-obtuse, and retuse at apex with a fine apiculus in the sinus. The flowers are fragrant, and apparently more long-lasting than those of other *Thrixspermum*-species. The inflorescences are axillary and many-flowered, but only one flower appears at a time on the thickened and laterally flattened rhachis, which is about 40 mm long.

Bracts: Triangular-apiculate, 9 mm long, and pale green, flushed with purple near apex.

Pedicel + ovary: 12 mm long.

Dorsal sepal: Narrowly elliptic-acute, 27 mm long, 10 mm wide.

Fig. 213. *Acampe rigida* (Buch.-Ham. ex J.E. Sm.) Hunt. August. 3.6×.

Lateral sepals: Slightly falcate, elliptic-acute, 28 mm long, 13 mm wide. All sepals white, spreading, rather thick.

Petals: Narrowly elliptic-acute, 21 mm long, 17 mm wide; slightly thickened, spreading, white.

Lip: Trilobed. Side lobes erect, small, pointed; curved towards apex of lip. Mid-lobe forming blunt, swollen, sac-like structure, 9 mm long, 7 mm wide. Base of lip saccate, enclosing column together with side lobes. Lip white, with diffuse orange to yellow band between side lobes and mid-lobe. Sac-like mid-lobe closed by pink-coloured septum, on top of which is a pale pink, tongue-like appendix, pointing upwards (3.5 mm long, 2.5 mm wide at base). Base of lip with triangular marking, widest at claw; yellow, with small red spots, and yellow glandular hairs.

Column: Straight, triangular, 9 mm long, 6 mm wide (at base), white.

Anther: White. Pollinia 4, in two groups of 1 large and 1 small pollinium.

Distribution: Thailand, Malaysia, Indonesia.

GENUS ACAMPE Lindley 1853

A small, but rather widely represented genus, comprising some 10 species distributed from Africa to Indochina, with one species in Malaya (only Penang and Langkawi). The generic name is derived from a Greek word meaning 'stiff', probably referring to the fleshy flowers and leaves in all the species.

Acampe rigida (Buch.-Ham. ex J.E. Sm.) P.F. Hunt, syn.: *Acampe longifolia* (Lindl.) Lindl., *Acampe penangiana* Ridl. [Fig. 213]

Sturdy epiphyte with monopodial growth. Stems more or less erect and elongated, completely covered by leaf-sheaths, leafy at apex only, with numerous, very sturdy (to 10 mm wide) aerial roots.

Leaves: To about 20 cm long, 2-6 cm wide, much thickened, leathery, oblong, broadly V-shaped in section; unequally bilobed at apex, lobes rounded (shorter lobe often more or less truncate). Sheaths about 2 cm long.

Inflorescences: Capitate, axillary, with about 8 flowers; peduncle to about 5 cm long, 3 mm wide, with minute purple spots, more densely spotted in lower part.

Bracts: Triangular-rounded with scarious apex; fleshy or swollen at base, 5 mm long and wide.

Flowers: Not resupinate; rather persistent (up to 3 weeks in cultivation, unpollinated); scented (scent reminding of *Galium verum*). All tepals greenish-yellow, fleshy, especially at apex, with transverse purple streaks, less well defined on rear side.

Pedicel + ovary: 13 mm, pale yellowish green.

Sepals: Obovate-rounded; 10 mm long, 5 mm wide; slightly incurved; 3 mm wide at base.

Petals: Spatulate-rounded; 10 mm long, 3 mm wide; slightly incurved, 1.5 mm wide at base.

Labellum: (Ligulate)-rounded, with lacerate margins; saccate, with deflexed apex; 10 mm long from onset of ovary; epichilium more or less triangular, papillose, white, with two transverse, purple streaks in proximal half. Base white, with erect margins ending in blunt points at mesochilium, and with a few purple spots; inside of basal, saccate part with dense, white, glandular hairs and a central, low callus.

Column: Short, with two pointed, subulate, decurved stelidia; greenish yellow, with purple spots; viscidium sharply upturned. Operculum greenish-yellow.

Fig. 214. *Cleisostoma subulatum* Bl. August. 0.4×.

Fig. 215. *Cleisostoma subulatum*. 5.7×.

Fig. 216. *Cleisostoma rostellatum* (Ridl.) Garay. August. 0.5×.

Fig. 217. *Cleisostoma rostellatum*. 8.4×.

Capsules: Elongated; to about 6 cm long, 0.9 cm wide.
Distribution: From India and Sri Lanka through Burma to Indochina and Hongkong.

GENUS CLEISOSTOMA Blume, 1825

A rather large epiphytic genus with more than 90 species in South Asia. Malaya has about ten species. The inflorescence has many, rather small, fleshy flowers, with a complicated structure. The lip is trilobed and spurred. At the back of the spur there is a large callus, obstructing more or less completely the entrance to the spur (Gk 'kleistos' = closed; 'stoma' = a mouth).

Cleisostoma subulatum Bl. [Figs 214 & 215]
This lowland epiphyte has pendulous and branching stems, slightly zig-zag and up to 70 cm long. The stems are provided with numerous aerial roots, and the internodes are between 2.5 and 3 cm long. Leaves with long sheaths; blades narrowly lanceolate to linear, acute, unequally and convergently bilobed at apex, and up to 35 cm long and 1.6 cm wide. There is a constriction of the lamina at 1 to 4 cm below the apex (commonly at 3-3.5 cm). There are often numerous, many-flowered inflorescences. These are pendulous and up to 19 cm long. The inflorescence sometimes has a side-branch, which can be up to 7.5 cm long.

Bracts: Short, scarious.
Pedicel + ovary: 6 mm long.
Dorsal sepal: Narrowly obovate-apiculate, slightly recurved, 5-6 mm long, 2.5-3 mm wide.
Lateral sepals: Obovate-apiculate, 6 mm long, 3 mm wide.
Petals: Oblanceolate-apiculate, 4-5 mm long, 1.5-2 mm wide. Sepals and petals mauve, with pale green margins and mid-vein.
Lip: Trilobed. Mid-lobe triangular-apiculate (or acuminate), 2.5-3.5 mm long, 2.5 mm wide; apiculus pointing upwards; white, flushed with very pale lilac. Side lobes curved, acuminate, pointing at each other; yellow.
Spur: Conical, curving slightly forward, 7 mm long; white, with faint lilac flush in front.
Column: White; operculum pale yellow.
Seed capsules: Pointing upward, at 45° with rhachis; 20 mm long, 5 mm wide; with many longitudinal grooves.
Distribution: India, Thailand, Vietnam, Malaysia, E. Indonesia.

Cleisostoma rostellatum (Ridl.) Garay [Figs 216 & 217]
A highland epiphyte with rather short stems, 4-15 cm long. The leaves are more or less tufted, 5 to 18 cm long and up to 2 cm wide, thick and leathery, almost oblong, and slightly tapering towards the apex, which is unequally, obtusely and divergently bilobed. The many-flowered inflorescence is up to 15 cm long, green, with purple streaks, and has one or a few, spreading, short branches. The flowers are persistent, and face the rhachis. Sepals and petals are spreading, and purple, with pale green base, margin and mid-vein. The small, triangular, and scarious bracts are 1.5 mm long, the pedicel + ovary 8 mm.

Dorsal sepal: Oblong-subacute (sometimes with a few, very small teeth at apex); 4.5-6 mm long, 2.5-3 mm wide; apex decurved.

Fig. 218. *Gastrochilus calceolaris* Don. Alt. 1600 m. June. 3.5×.

Fig. 219. *Gastrochilus calceolaris*. 10.5×.

Lateral sepals: Oblong-obtuse (or with very small teeth at apex); 5 mm long, 1.2 mm wide.

Lip: Trilobed. Mid-lobe triangular or nearly hastate; acute, swollen, porrect, white, 3.5 mm long. Side lobes obliquely truncate, erect, 2 mm long; yellow; rear edge purple, with small, white, tooth-like processes at base. Base of lip saccate, length of sac 5.5 mm from base of lateral sepals. Sac deflexed, white; anterior wall with low septum, starting on mid-lobe. Posterior wall with pale yellow, bifid, pubescent callus.

Column: With wide, winged base; apex with two falcate processes. Whole column pale green, flushed with purple. Rostellum bifid. Operculum white.

Seed capsules: To 16 mm long.

Distribution: Malaysia, Indonesia.

GENUS GASTROCHILUS D. Don, 1825

This is a small epiphytic genus with only two species in Malaya. These are rather small and short-stemmed, with a few leaves. The flowers are not very large, but quite attractive and distinctive, having a large, rounded sac at the base of the lip (Gk 'gaster' = a belly; 'kheilos' = a lip), and a small, flat mid-lobe near the rim of the sac. The side lobes are erect, or connivent across the base of the mid-lobe.

Gastrochilus calceolaris Don [Plate 5.4; Figs 218 & 219]

A small highland epiphyte with a short stem, 0.5-6 cm long. The leaves are obliquely lanceolate, 3-12 cm long, 0.6-1.8 cm wide, rather fleshy and spotted with purple. The lower side is often flushed with purple. They are very distinctly and subacutely bilobed, although the lobes are sometimes convergent and less distinct. One lobe is always longer than the other. One plant can have up to five inflorescences, each with 1-7 flowers. The flowers are not resupinate, and have minute, amplexicaul bracts.

Pedicel + ovary: 6 mm long.

Sepals and petals: Spreading, spatulate, 7 mm long, 2-3 mm wide; yellow-green, with purple spots. Dorsal sepal convex, with incurved apex.

Lip: Forming large, 6-7 mm long sac; yellow, with purple spots. Claw wide, attached to wings and base of column; whitish. Small white callus at juncture of side- and front lobes. Apex of lip triangular, deeply fringed with white hairs; erect; 4-5 mm wide, to 3 mm long; with yellow streak or triangle, and purple spots in centre.

Column: Mauve, turning very pale at apex. Operculum yellow or pale green. Rostellum very long, apex folded over backwards.

Capsule: 20 mm long, 6 mm wide.

Distribution: Nepal, Bhutan, India, China, Burma, Thailand, Vietnam, Malaya.

GENUS SCHOENORCHIS Reinwardt (or Blume, 1825)

This small genus is widely distributed through South Asia, with three species in Malaya. Two of these are rooting at the base only, like sympodial orchids. The inflorescence is many-flowered, with minute flowers that do not open much. The lip is spurred, and the anther is provided with a long, up-turned process.

Fig. 221. *Schoenorchis minutiflora*. 8.0×

Fig. 220. *Schoenorchis minutiflora* (Ridl.) J.J. Sm. Alt. 1340 m. November. 1.1×.

Fig. 222. *Uncifera tenuicaulis* (Hook. f.) Holttum. November. 1.1×.

Fig. 223. *Uncifera tenuicaulis*. 7.5×.

Schoenorchis minutiflora (Ridl.) J.J. Sm. [Figs 220 & 221]
A mountain epiphyte with short, erect stem, 8-15 cm long; internodes 0.3-1 cm; leafy throughout; branching at the base.
Leaves: Lanceolate, unequally bilobed at apex; thickened, V-shaped in section; to 6 cm long, 0.5-1 cm wide.
Inflorescences: Axillary, much branched; arching so that main branch is nearly at 180° with base of scape. Main branch 4-20 cm long. Side branches about four, 2-10 cm long. Inflorescence with many, very small, white flowers, 2 mm long from apex of lip to base of ovary. Flowers start opening at base of main branch, opening much later on lateral branches. Only a few flowers open at a time. Scape to 8 cm long, distinctly flattened (2.5 × 1 mm), with two tightly clasping, dark brown, ovate sheaths, one immediately below first side branch. Subsequent lateral branches with sheaths getting gradually smaller.
Bracts: Floral bracts triangular-acute, dark brown, about 1 mm long.
Pedicel + ovary: 1-1.5 mm long.
Sepals: Ovate-rounded; dorsal sepal 1 mm long; laterals broadly ovate, slightly oblique, 1.5 mm long.
Petals: Oblong-truncate, 0.5 mm long.
Lip: Trilobed, spurred, 3 mm long. Mid-lobe more than 1 mm long, thickened, rounded. Side lobes erect, rounded.
Spur: Conical with blunt apex; parallel to ovary; 1.2 mm long.
Column: Green; operculum yellow.
Distribution: Thailand, Malaya.

GENUS UNCIFERA Lindley, 1859

A small South- and East Asian genus, with only one species in Malaya. Most species, including the Malayan one, have a long, hanging stem with many leaves. The inflorescence is very short, with a few flowers only. These have a short, swollen spur which is curved forwards. The stipes of the pollinia is rather conspicuous, being long, widened at the apex, and having a large viscidium. Etymology: L. 'uncus' = a hook; '-fer' = carrying.

Uncifera tenuicaulis (Hook. f.) Holttum [Plate 5.5; Figs 222 & 223]
A lowland epiphyte with slender, pendulous stems (2 mm wide), about 40 cm long, and sturdy aerial roots (2.8 mm wide); leafy throughout.
Leaves: Linear-acute, thin but stiff; to 17 cm long, 0.6 cm wide; with marked mid-vein and a constriction in apical part.
Inflorescences: Very short-stemmed, with 1-3 persistent flowers.
Bracts: Floral bracts minute, triangular, scarious.
Pedicel + ovary: 3 mm long; white, with longitudinal purple stripes.
Dorsal sepal: Obovate-rounded, 3 mm long, 1.5 mm wide, much convex, forming hood; apex slightly upturned.
Lateral sepals: Obliquely obovate-rounded, spreading, 4 mm long, 2 mm wide.
Petals: Ovate-subacute, 3 mm long, 1.5 mm wide. All sepals and petals yellow, with diffuse purple streak or spot in centre.
Lip: Trilobed, 5 mm long. Mid-lobe forming narrow tube, 0.6 mm wide, widening to 1 mm below apex. Apex bifurcate, with two falcate, upturned, white points; rest of

Fig. 224. *Pteroceras appendiculatum* (Bl.) Holttum. Alt. 800 m. August.

Fig. 225. *Pteroceras appendiculatum*. 5.8×.

Fig. 226. *Pteroceras appendiculatum*. 5.7×.

mid-lobe mauve. Side lobes erect, with straight, parallel sides; apex uneven. Base of lip with sac-like spur, 3 mm long from onset of ovary, 2 mm wide; very pale yellow.

Column: Short, thick, pale green; rostellum rounded and upturned. Operculum yellow, pointing upwards. Stipes much widened towards apex. Viscidium large, ovoid, pale green.

GENUS PTEROCERAS Hasskarl, 1842

A South Asian epiphytic genus with about 40 species, previously referred to the (Australian) genus *Sarcochilus*. The Malayan species number about 15. The plants have short stems and fleshy leaves. Sepals and petals are similar, and spreading widely. The column is short, but has a long foot, to which the lateral sepals are joined. The lip is hinged to the end of, and in line with, the column-foot. It is saccate (or spurred) and trilobed, the mid-lobe being thickened and rather short. The side lobes are distinct, curved and erect. There are two, partly and unequally cleft pollinia. Etymology: Gk 'pteron' = a wing; 'keras' = a horn.

Pteroceras appendiculatum (Bl.) Holttum, syn. *Grosourdya appendiculata* (Bl.) Reichenb. f. [Figs 224, 225 & 226]

A small lowland epiphyte with a very short stem, less than 1 cm long.

Leaves: The lanceolate leaves are close together and more or less in one plane; lanceolate, falcate; acutely, unequally and connivently bilobed at apex; 10 cm long, 1.8 cm wide; rather thick and stiff, with depressed mid-vein.

Inflorescences: 1-4, scape 3-5 cm long, deep purple, densely hairy, with one spreading stem bract. Rhachis short, slightly thickened; with acute, spreading bracts, 1 mm long. Pedicel + ovary 12 mm long. One flower open at a time on many-flowered inflorescence. When the flower is fully open, all tepals are sharply reflexed, parallel to ovary. Flowers last one day, at the end of which tepals bend forward again. All tepals similar; pale yellow, with reddish purple spots.

Dorsal sepal: Narrowly oblanceolate-acute, apex upturned; 8.5 mm long, 2.5 mm wide.

Lateral sepals: Narrowly and obliquely elliptical-acute, 8 mm long, 4 mm wide; joined to base of column-foot.

Petals: Narrowly oblanceolate-acute, 7 mm long, 2.2 mm wide.

Lip: Trilobed, in line with column-foot, base joined to column-foot. Mid-lobe with short central white point, and two ligulate, oblique, yellow lobes, 2 mm long, pointing slightly backwards. Blade of lip white. Side lobes subulate, erect, with back-turned apices; 3 mm long, pale mauve; base yellow. Base of lip with saccate spur, pointing forwards; 3 mm long, 2 mm wide at apex; yellow. Mouth of spur with erect, tubular keels.

Column: Pale yellow, slender, widening towards apex; with long, slender, purple column-foot. Rostellum long, pointed. Pollinia on long, oblanceolate stipes with triangular viscidium. Operculum white, with long point.

Distribution: Andaman Isl., Burma, Vietnam, Thailand, Malaysia, Philippines.

Fig. 227. *Phalaenopsis violacea* Witte. Courtesy of Royal Botanic Gardens. Kew.

Fig. 228. *Ascochilopsis myosurus* (Ridl.) Carr. October. 1.5×.

GENUS PHALAENOPSIS Blume, 1825

A rather large genus of considerable horticultural interest, including the lovely *Phalaenopsis amabilis* with large, white flowers, which is found on the Sunda Islands and in the Philippines. The Malayan species number about five, two of which are commonly seen in cultivation: *P. violacea* and *P. cornucervi*. The name of the genus means 'looking like a moth', and the species *P. amabilis* referred to above is often called 'Moth Orchid'. A characteristic feature of the genus is the forked appendage on the lip.

Phalaenopsis violacea Witte [Fig. 227]

An attractive lowland epiphyte, mainly found in shady places in riverine forest (Perak and Selangor). Stem very short.

Leaves: The closely tufted and shiny dark green leaves are obovate-obtuse, to about 25 cm long, 8 cm wide, thickened and coriaceous.

Inflorescences: 1-5, slender, rather short; rhachis slightly zig-zag; flattened; 2-5-flowered. Flowers to about 7 cm across, rather fleshy and persistent.

Bracts: Alternate, 6 mm long, pale green.

Sepals: Ovate-acute; to 25 mm long, 10 mm wide; purple, with pale green apices and (often) margins; much spreading; laterals slightly falcate.

Petals: (Ob)ovate-acute; as large as sepals, or slightly shorter; pale green, flushed with purple.

Lip: Trilobed, to 20 mm long. Mid-lobe to about 10 mm long, 7.5 mm wide; more or less orbicular-acute; purple, thickened; with short apiculus, pointing downward, below apex. Side lobes 7 mm long, erect, anterior margins incurved, touching; yellow. Blade of lip between side lobes with two narrow tooth-like appendages; yellow; pointing towards apex.

Column: Slender, purple, decurved.

Distribution: Malaysia, Sumatra.

GENUS ASCOCHILOPSIS Carr, 1929

A South Asian genus with only one species: a small lowland epiphyte with a characteristic, swollen and scurfy rhachis. The flowers are small and yellow. The name of the genus means 'similar to *Ascochilus*', another genus of orchids.

Ascochilopsis myosurus (Ridl.) Carr [Figs 228 & 229]

A lowland epiphyte with a very short stem, 3-7 cm long. The thickened leaves are more or less obliquely lanceolate-acute, 10 cm long, 1.3 cm wide, unequally and conniventy bilobed at apex, and with a depressed mid-vein. The scape is slender, 4-5 cm long and 1 mm wide, with short, thick, greyish hairs and one minute, acute stem bract. Rhachis swollen, 2.5 mm wide, 0.5-4 cm long, club-shaped and many-flowered, but only a few flowers are open at a time. Flowers very shortly persistent.

Bracts: Triangular-acute, 5 mm long.

Tepals: All similar, about 3 mm long, slightly spreading, yellow.

Dorsal sepal: Obovate-rounded, 1.5 mm wide.

Petals: Oblong-rounded, 0.8 mm wide.

Lip: Much swollen, adnate to base of column; spurred; white. Side lobes pointing

Fig. 229. *Ascochilopsis myosurus*. 7.5×.

forward, longer than mid-lobe, close together but with slightly diverging apices; yellow marking on inner side. Mid-lobe triangular-obtuse, with wide base. Whole lip 3 mm long, 1.5 mm wide.

Distribution: Malaya, Sumatra.

GENUS VANDA W. Jones ex R. Brown, 1795

A genus of great horticultural interest, numbering some 50 species in South Asia (from India and Sri Lanka through Malaysia to the Philippines). The plants in this genus often have stout, elongated stems, very thick roots, and leaves at an acute angle to the stem. The persistent flowers are generally rather large and fleshy. Sepals and petals are nearly equal, narrowing towards the base, often tessellated and with more or less reflexed and wavy edges. The 3-lobed lip is unmovably attached to the very short column-foot. Two species have been recorded for Malaya. In addition, Vanda hybrids are commonly grown as ornamental plants in gardens throughout the tropical parts of Asia.

Vanda hookeriana Reichenb. f., syn. *Papilionanthe hookeriana* (Reichenb. f.) Schltr

Stems long and slender, climbing or scrambling.

Leaves: Terete, thickened, about 10 cm long, 3 mm wide, with a constriction about 2 cm from the apex.

Inflorescence: To 20 cm long, sparsely many-flowered.

Tepals: About 2 cm long, 1.5 cm wide; white or very pale mauve.

Lip: Deeply three-lobed; side lobes spreading, deep purple; mid-lobe broad, pale mauve with deep purple markings. Found in Johore, Terengganu and Perak (Kinta Valley: this species is sometimes referred to as 'Kinta Weed').

Distribution: Thailand, Malaysia, Indonesia, Vietnam.

Botanical glossary (with special reference to orchids)

Accrescent: getting longer or thicker.
Accumbent: lying against another part.
Acicular: with a very narrow and stiff point.
Acuminate: with a long, tapering point.
Acute: ending in a sharp point.
Adnate: grown together, fused with (said of different parts).
Adpressed: pressed close to.
Adventitious: occurring in an abnormal position.
Alternate (said of leaves, etc.): placed at different levels along stem, cf. 'opposite'.
Amplexicaul: clasping the stem.
Ampulliform: flask-shaped.
Androclinium: a cavity at the apex of the column where the anthers rest in many orchids.
Anterior: facing away from the axis.
Anther: the part of the stamen that contains pollen.
Anther canal: antheric process.
Anther cap (= operculum): anther modified into little lid, covering pollinia.
Anticous: facing inward toward the axis, or in the position furthest from the axis.
Annulate: ringed; surrounded by rings.
Apex, pl. apices: the end farthest from the point of attachment.
Apiculate: ending in a small point.
Apiculus, pl. apiculi: a small point.
Appendage: a subordinate or external part.
Appressed: see 'adpressed'.
Arcuate: curving like a bow.
Aristate: with a long bristle or bristles.
Assimilate: to absorb and build up foodstuffs into complex constituents of an organism.
Auricle: an ear-like part.
Auriculate: with an ear-like part (lobe, appendage, etc.).
Axil: the angle between the upper side of an organ, e.g. a leaf, and an axis, e.g. a stem.
Axillary: arising in the axil.
Barbed: provided with stiff bristles, pointing backwards.
Basal: at the lower end.
Bearded: with long hairs.
Bifid: split into two parts at the apex.
Bifurcate: forked, divided into two parts.
Bilobed: with two lobes.
Bipartite: divided into two parts nearly to the base.
Botuliform: sausage-shaped.
Bract: modified leaf at the base of a flower or of a branch of an inflorescence.
Bracteate: provided with bracts.
Bracteole: a small bract.
Bullate (said of a leaf, petal, etc.): having inflated areas between the veins.
Caducous: falling off early.
Caespitose: forming mats.
Calcar: a spur.
Callus, pl. calli: a hardened projection.
Calyptriform: cap- or hood-like.
Calyx: the outermost part of a flower; the sepals.

Campanulate: bell-shaped.
Canaliculate: with a longitudinal groove.
Canescent: greyish white or hoary.
Capitate: growing in a head, or with an enlarged head.
Capsule: dry fruit with many seeds in more than one section, splitting open when ripe (in the case of orchids).
Carpel: female reproductive organ of flowering plants, consisting of ovary and stigma.
Caruncle: a process near the hilum of a seed.
Caudate: with a slender, tail-like appendage.
Caudicle: a pollen stalk.
Caulescent: provided with a stem.
Chaffy: covered with, or resembling, light dry scales.
Chartaceous: with a paperlike texture.
Ciliate: with small hairs forming a fringe.
Ciliolate: with a fringe of very small hairs.
Clavate: club-shaped.
Claw: narrow, basal part of petal, especially of lip.
Cleistogamous: with unopened, self-pollinating flowers.
Clinandrium: see Androclinium.
Cohere: to be connected.
Collateral: side by side, parallel.
Column: union of stamens and style in orchids.
Complicate: folded lengthwise.
Concave: hollow, with a surface that curves inward.
Conduplicate: folded together lengthwise.
Confluent: united, grown together.
Connate: firmly united.
Connivent: with the ends inclined towards each other.
Conspicuous: easily seen, attracting notice.
Contiguous: close together, touching, bordering.
Convex: with a surface that curves outwards.
Convolute: rolled up, coiled.
Cordate: heart-shaped, and wide at base.
Coriaceous: leathery.
Corolla: collective name for the petals in a flower.
Corrugate: with parallel ridges and furrows.
Crateriform: bowl-shaped.
Crenate: with rounded teeth.
Crenulate: with very small teeth.
Crispate: curled.
Cucullate: hood-like (conical).
Cuneate: wedge-shaped (especially leaf-base).
Cupular: cup-shaped.
Cusp: a rigid, sharp point.
Cuspidate: having a rigid, sharp point.
Cymbiform: boat-shaped.
Cyme: a branched inflorescence where each main and secondary stem bears a single flower, the central flower opening first.
Deciduous: shedding leaves at a certain season or stage of growth.
Declinate: bent downward or aside.
Decurrent (of leaf): with base extending downward along the stem.
Decurved: curved downwards.
Deflexed: bent downwards.
Deltoid: shaped like the Greek letter delta, Δ.
Dentate: with pointed, outward directed teeth.
Denticulate: finely toothed.
Depressed: flattened vertically.

Dimidiate: halved, split on one side.
Discoid: like a disk or plate.
Distal: situated away from point of attachment.
Distichous: placed one above the other in two regular, vertical, opposite rows.
Divaricate: branch off, spread out widely.
Divergent: going in different directions.
Dorsal: of, on, or near the back.
Elongate(d): long and narrow.
Emarginate: with a notch at the tip.
Embryo: the rudimentary plant in a seed.
Endemic: confined to a particular area.
Endosperm: nourishment for embryo, surrounding it in a seed.
Ensiform: having the shape of a sword.
Entire: without teeth or lobes.
Epichilium (= epichile): front part of lip.
Epiphyte: plant, using another plant for support without obtaining food from it.
Equitant: overlapping.
Erect: directed upward.
Erose: with small, irregular notches in the margin.
Exserted: sticking out.
Falcate: curved like a sickle.
Farinose: covered with a white, mealy substance.
Fascicle: a tuft or cluster of leaves or flowers.
Fasciculate: growing in clusters or bundles.
Filiform: thread-like, very thin.
Fimbriate: fringed (with thicker processes than hairs).
Flabellate: fan-shaped.
Flabelliform: fan-shaped.
Flexuous: bent alternately from joint to joint, from flower to flower, etc.; zig-zag.
Floccose: covered with wooly hairs.
Foliaceous: leaf-like.
Fornicate (of leaf, petal, etc.): arching over another part.
Fugacious: said of a part that is soon shed.
Fused: joined together.
Fusiform: spindle-shaped.
Galeate: helmet-shaped.
Geniculate: bent like a knee.
Germinate: to sprout, to start growing.
Gibbose (= gibbous): rounded and bulging; pouched.
Glabrous: hairless.
Glandular: with small warty or fleshy processes.
Glomerate: in a dense cluster.
Granular: resembling grains; granulated.
Gregarious (of flowering): when flowering tends to take place simultaneously in individual plants of the same species, growing together in the same locality.
Gynaecium: the female organ of a flower.
Habitat: a particular environment inhabited by a plant.
Hastate: shaped like a halberd, triangular with spreading basal angles.
Hibernate: to pass the dry season in a dormant condition.
Hilum: scar left on a seed from the attachment to the placenta.
Hirsute: with coarse hairs.
Hispid: bristly.
Hispidulous: having short, small bristles.
Hyaline: thin and translucent; glassy.
Hybrid: result of crossing of different species (or sometimes genera).
Hypochilium (= hypochile): the basal part of a lip.
Hysteranthous: putting forth leaves after the flowers have opened.

Imbricate: overlapping, like roof-tiles.
Incised: deeply cut.
Incumbent: leaning on or resting against.
Incurved: curved inward.
Indehiscent (of fruit): not opening at maturity.
Indumentum: a hairy (wooly, scaly) covering.
Induplicate: having the edges folded in, but not overlapping.
Inferior: located, or growing, below another part.
Inflated: blown up.
Inflexed: bent inward.
Inflorescence: flowering shoot.
Infundibular: funnel-shaped.
Internode: part of stem between successive nodes.
Involucre: a ring of small leaves or bracts at the base of a flower or a flower cluster.
Involute: rolled inward at the edges.
Keel: a sharp ridge.
Labellum: lip.
Lacerate: with an irregular edge, as if torn.
Laciniate: cut into slender lobes.
Lamella, pl. lamellae: a thin plate.
Lamellate: consisting of, resembling, or provided with little plates.
Lamina, pl. laminae: upper, spreading part of tepal; a leaf-blade.
Lanceolate: narrow and gradually tapering toward apex.
Lateral: on, at, of, etc. the side.
Latex: a milky juice produced by plants.
Lax: (of inflorescence) loose, open.
Lenticular: lens-shaped (with two opposing, convex sides).
Lepidote: covered with scales.
Ligulate: strap- or tongue-shaped.
Ligule: a thin strap-shaped process.
Linear: long, narrow and with two parallel sides.
Lip: the lower or middle petal of an orchid.
Lithophyte: a plant that grows on rock.
Lobe: a (rounded) projection or division.
Lobulate: with small lobes.
Loculus, pl. loculi: a small cavity or compartment in plant tissue.
Lorate, loriform: strap-shaped.
Lunate: formed like a half-moon.
Mammillate: provided with nipples (mammillae).
Mentum: projection in front of flower, formed by column-foot and lateral sepals.
Mesochilium (= mesochile): middle part of lip.
Mitriform: mitre-shaped.
Monopodial: having single main axis, continuing to grow at apex.
Mucronate: ending abruptly in a 'mucro', or sharp point.
Muticous: pointless, blunt.
Mycorrhiza: an association of a fungus with a root of a higher plant.
Nectar: a sugary liquid attractive to insects.
Nectarium (= nectary): part of plant secreting nectar.
Node: point of stem from where one or more leaves or branches arise.
Obcordate: heart-shaped, with the broad end upward.
Oblanceolate: narrow and gradually widening toward apex.
Oblique: with unequal sides.
Oblong: elliptical, and two to four times longer than wide.
Obovate: egg-shaped, with the broader side upward.
Obtuse: blunt.
Operculate: with an anther cap.
Operculum: anther cap q.v.

Opposite (leaves): placed at the same node on opposite sides of stem.
Orbicular: round and flat.
Ovary: hollow lower part of carpel containing the immature seeds.
Ovate: egg-shaped, with the broader end at the base.
Paleaceous: chaff-like.
Panduriform: shaped somewhat like a violin.
Panicle: an irregularly branched inflorescence.
Paniculate: with a panicle.
Papilla, pl. papillae: a tiny protruding cell.
Papillose: covered with papillae.
Parasite: a plant, growing on another plant, and obtaining food from it.
Partite: deeply divided into parts.
Pectinate: comb-like.
Pedicel: a stalk of an individual flower in an inflorescence.
Peduncle: a stalk of an inflorescence or of a solitary flower.
Pellucid: transparent, translucent.
Peltate (of leaf, etc.): having the stem inserted into the lower surface of the lamina, not far from the centre.
Pendulous: hanging down.
Penicillate: having, or resembling, a tuft of hairs.
Perianth: collective term for petals + sepals.
Persistent: remaining attached for a long time.
Petiole: a stalk of a leaf.
Pilose: covered with fine, soft hair.
Pinnate: with leaflets on both sides of a common stem in a feather-like arrangement.
Pisiform: resembling a pea.
Plicate: folded like a fan.
Pollinate: to convey pollen from an anther to a stigma.
Pollinium, pl. pollinia: pollen grains clumped together.
Porcate: ridged.
Porrect: extended horizontally.
Posterior: near the axis.
Praemorse (= premorse): looking as if bitten off.
Precocious: developing early.
Process: a projecting part of an organism.
Prostrate: lying flat (on the ground or on the substrate).
Proteranthous: bearing flowers before leaves.
Proximal: near the place of attachment.
Pruinose: covered with a white, hoarfrost-like bloom.
Pseudobulb: swollen part of stem (internode), often resembling a bulb.
Puberulent: very finely pubescent.
Pubescent: covered with soft, downy hairs.
Pulverulent: looking as if covered with powder.
Pulvinate: cushion-shaped.
Punctate: marked with spots.
Punctulate: marked with small spots.
Pungent: ending in a hard, sharp point, or with an acrid smell.
Pustulate: covered with small blisters.
Pyriform: shaped like a pear.
Quadrate: square or nearly square.
Raceme: inflorescence with elongated axis, carrying simple pedicels.
Racemose: in the form of a raceme.
Receptacle: the extremity of the stalk on which the flower grows, the torus.
Recurved: curved downward and/or backward.
Reflexed: bent downward and/or backward.
Reniform: kidney-shaped.
Repand: having an uneven, slightly wavy margin.

Resupinate (of flower): having lip downward and dorsal sepal upward.
Reticulate: having a structure like the meshes of a net.
Retrorse: bent abruptly backwards.
Retuse: with a rounded end, the centre of which is depressed.
Revolute: curled or rolled backward or downward at margins or tip.
Rhachis (= rachis): the principal axis of an inflorescence.
Rhizome: a root-stock.
Rostellum: an elevated projection of the stigma in orchids.
Rostrate: with a beak-like process.
Rostrum: a beak-like process.
Rotund: orbicular, round, without angles.
Rugose: wrinkled.
Saccate: with, or having the form of, a pouch or sac.
Scabrid: slightly roughened.
Scabrous: rough with small points.
Scandent: climbing.
Scape: a leafless flower-stalk, growing from the crown of a root.
Scapigerous: with, or looking like, a scape.
Scarious: thin, dry, and semi-transparent.
Scurfy: covered with little scales.
Secund: with parts arranged on one side only (e.g. flowers in an inflorescence).
Septate: with a dividing wall.
Septum: a dividing wall, a partition.
Sericeous: covered with short, straight, adpressed silky hairs.
Serrate (of leaf): with sharp notches along the margins, pointing toward the apex.
Serrulate: finely serrate.
Sessile: without a stalk.
Setaceous: bristly.
Setose: bristly.
Sigmoid: curved like the letter S.
Sinuate: with a wavy margin (rather deeply so).
Spat(h)ulate: spoon-shaped (but often quite flat).
Spicate: arranged in a spike.
Spiciform: spike-like.
Spike: inflorescence with flowers attached directly to an undivided rhachis.
Staminode: a sterile, rudimentary stamen.
Stigma: the surface of the pistil which receives the pollen.
Stipe: the stem supporting a pistil.
Stipitate: having a stipe.
Stria, pl. striae: one of a number of fine, (parallel) streaks or shallow furrows.
Striate: provided with striae.
Style: the part of a pistil between the stigma and the ovary.
Sub-: a prefix, meaning under, below, to a lesser degree than, somewhat, slightly, etc.
Suberose: 1. looking as if slightly eaten; 2. corklike.
Subulate: awl-shaped.
Sulcate: with deep, parallel furrows or grooves.
Superior: located, or growing, above another part.
Superposed: placed on, over, or above something else.
Sympodial: having a growth with yearly lateral shoots.
Synanthous: having the flowers and leaves appearing at the same time.
Synsepalum: structure formed by the union of lateral sepals.
Taper: to decrease gradually in width.
Tepal: sepal or petal (generally excluding lip).
Terete: circular in cross-section.
Terminal: at the end of a stem.
Terrestrial: growing in the ground.
Tessellate(d): checkered.

Testa: outer coat of a seed.
Tetragonal: (of stem) with four (obtuse) longitudinal angles.
Tetragonous = tetragonal.
Tetraquetrous: (of stem) with four very acute longitudinal angles (wings).
Tomentose: having dense, short, soft hairs.
Tomentum: mass of long, soft entangled hairs, adpressed to surface.
Tortuous: twisted.
Torus: see Receptacle.
Translucent: letting light pass through.
Transverse: placed across.
Trigonal: (of stem) with three (obtuse) longitudinal angles.
Trigonous = trigonal.
Trilobed: with three lobes.
Triquetrous: (of stem) with three very acute, longitudinal angles (wings).
Truncate: appearing as if abruptly cut off at the tip.
Tuber: swollen part of underground stem, with dormant buds. Tubers serve as storing organs for nutrients.
Tubular: tube- or pipe-like.
Tumid: swollen.
Turbinate: top-shaped, with a narrow base and a wide apex.
Turgid: swollen, inflated.
Umbel: inflorescence with divergent pedicels arising in the same point.
Umbonate: with a small central elevation.
Uncinate: with a hook or barb at the end.
Undulate: having a wavy margin.
Unguis: the claw-shaped base of a petal.
Urceolate: urn-shaped.
Vegetative: relating to the parts of a plant concerned with growth, as opposed to those concerned with reproduction.
Velamen: water-absorbing outer cover of aerial root, consisting of several layers of dead cells.
Velutinous: velvet-like.
Venation: collective term for veins (in a leaf, etc.).
Venose (= venous): having veins.
Ventral: of, or belonging to, the inner face.
Ventricose: bulging out on one side.
Vermiform: worm-like.
Verrucose: warty, or wart-like.
Versatile: said of part which is freely movable on the filament to which it is attached.
Villous: covered with long, soft, fine hairs.
Viscidium: part of rostellum forming a sticky disc, to which pollinia are attached.
Whorl: arrangement of leaves, etc. in a circle.
Wing: flattened extension of stem, etc.
Zygomorphic: bilaterally symmetrical.

References

Chin, Hoong Fong 1977. *Malaysian Flowers in Colour*. Tropical Press Sdn. Bhd., Kuala Lumpur.

Darwin, Charles 1984. *The Various Contrivances by which Orchids are Fertilised by Insects*, 2nd edition. The University of Chicago Press, Chicago and London.

Henderson, M.R. 1974 (reprint). *Malayan Wild Flowers Monocotyledons*. The Malayan Nature Society, Kuala Lumpur.

Henderson, M.R. & G.H. Addison 1969. *Malayan Orchid Hybrids*. Government Printing Office, Singapore.

Holttum, R.E. 1972. *A Revised Flora of Malaya*; *Orchids of Malaya* Vol. I, reprint of 3rd edition. Government Printing Office, Singapore.

Jayaweera, D.M.A. 1981. Orchidaceae. In: M.D. Dassanayake & F.R. Fosberg (editors), *A Revised Handbook to the Flora of Ceylon*, Vol. 2. Balkema, Rotterdam.

Millar, Andrée 1978. *Orchids of Papua New Guinea*. Australian National University Press, Canberra.

Seidenfaden, Gunnar 1976a. Contributions to the Orchid Flora of Thailand VII. *Botanisk Tidsskrift* 71, Bind 1-2. Haefte, Copenhagen.

Seidenfaden, Gunnar 1976b. Orchid Genera in Thailand IV. Liparis L.C. Rich. *Dansk Botanisk Arkiv*, Bind 31, Nr 1. Copenhagen.

Seidenfaden, Gunnar 1977. Orchid Genera in Thailand V. Orchidoideae. *Dansk Botanisk Arkiv*, Bind 31, Nr 3. Copenhagen.

Seidenfaden, Gunnar 1978. Orchid Genera in Thailand VI. Neottioideae Lindl. *Dansk Botanisk Arkiv*, Bind 32, Nr 2. Copenhagen.

Seidenfaden, Gunnar 1982. Orchid Genera in Thailand X. *Trichotosia* Bl. and *Eria* Lindl. *Opera Botanica* 62. Copenhagen.

Seidenfaden, Gunnar 1983. Orchid Genera in Thailand XI. Cymbidieae Pfitz. *Opera Botanica* 72. Copenhagen.

Seidenfaden, Gunnar 1985. Orchid Genera in Thailand XII. *Dendrobium* Sw. *Opera Botanica* 83. Copenhagen.

Seidenfaden, Gunnar 1986. Orchid Genera in Thailand XIII. Thirty-three epidendroid genera. *Opera Botanica* 89. Copenhagen.

Seidenfaden, Gunnar 1988. Orchid Genera in Thailand XIV. Fifty-nine vandoid genera. *Opera Botanica* 95. Copenhagen.

Veitch, James & Sons 1893. A Manual of Orchidaceous Plants Cultivated under Glass in Great Britain. Royal Exotic Nursery, Chelsea.

Index to orchids in the text

Malayan orchid genera are in CAPITALS, synonyms are in *italics*